Women and Ideas in Engineering

Women and Ideas in Engineering

TWELVE STORIES FROM ILLINOIS

LAURA D. HAHN AND ANGELA S. WOLTERS

UNIVERSITY OF
ILLINOIS PRESS
Urbana, Chicago, and Springfield

Library of Congress Cataloging-in-Publication Data
Names: Hahn, Laura D., author. | Wolters, Angela S., author.
Title: Women and ideas in engineering: twelve stories from Illinois /
    Laura D. Hahn and Angela S. Wolters.
Description: Urbana: The University of Illinois, [2018] | Includes
    bibliographical references and index.
Identifiers: LCCN 2018000542 | ISBN 9780252041969 (cloth : alk.
    paper)
Subjects: LCSH: Women engineers—United States—Biography. |
    Engineers—United States—Biography. | Illinois—Biography. |
    Engineering—Illinois—History.
Classification: LCC TA139 .H2734 2018 | DDC 620.0092/5209773—dc23
LC record available at https://lccn.loc.gov/2018000542
Ebook ISBN 978-0-252-05067-1

CONTENTS

FOREWORD

Today, we find ourselves at a pivotal point in human history. The past two hundred and fifty-odd years have been marked by some of the most spectacular scientific discoveries of humankind. Over the same period, the transformative breakthroughs in technological innovation that defined the past three industrial revolutions are enabling and inspiring us to tackle and conquer some of the world's most humbling challenges. Curing disease, eradicating famine, and expanding our living boundaries beyond the gravity of our home planet promise to mark the twenty-first century as one of the most brilliant eras in the history of our species.

Imagine how much further ahead we could have been if as many women as men had been allowed to immerse themselves in the discoveries and innovations of civilization.

The absence of women from the long list of the ancient Greek philosophers stands in striking contrast to the Greeks' worship of Athena as the Olympian goddess of wisdom, reason, mathematics, the arts, and many other of those things that make humans humane. Although expectations, traditions, arrogance, and ignorance have discouraged women from partaking in intellectual pursuits throughout history, there were those who engaged, who found a way to be involved and make an impact. As recently as the mid–nineteenth century, women were becoming engineers. As early as the early eighteenth century, *The Ladies' Diary*, an almanac including puzzles and challenging mathematical questions and their answers, enjoyed an early and long run, beginning in 1704. Imagine if, 3000 years ago, women, in full force, had been welcomed and encouraged to join the club of discovery and innovation. Imagine what the impact of the University of Illinois at Urbana-Champaign could have been if as many women as men had been part of the scientific inquiries and engineering pursuits that established it as one of the most influential land-grant universities of the nation.

Women sense expectations as well as prejudice. They like math until they are told to do other things. The list of reasons why women have not been equal participants in STEM can be reduced to one barrier, exclusion. Past inequalities are lamentable. However, twenty-first-century ideals of access and opportunity to education are understood to be an indisputable birthright, regardless of gender. Today, we cannot afford prejudice and we have great expectations.

This book is much more than a survey of women in engineering at the University of Illinois and their contributions and impact over the years. It reminds us of our duty to break down the barrier of exclusion. It is a glimpse at the bright future that lies ahead when the expectation is the fulfillment of potential, with men and women sharing in full force the responsibility to advance the world through scientific discovery and engineering innovation. It is a call to action for all of us at the University of Illinois to embrace the promise of the possibilities of all of our students. Barriers to the participation of women continue to be broken. The documentation of women emerging in STEM fields at the University of Illinois will inspire us to make sure that the desire to innovate, create, and discover will be respected as the gender-neutral pursuit that it must be, for the sake of progress and self-respect.

Andreas Cangellaris
Vice Chancellor for Academic Affairs and Provost,
University of Illinois at Urbana-Champaign
Dean, College of Engineering, 2013–2018,
M. E. Van Valkenburg Professor in Electrical
and Computer Engineering

PREFACE

Women in engineering are out there and on the rise: innovating, designing, building, researching, teaching, leading, and changing the world. Their contributions and their perspectives represent a growing force that educators and employers are coming to value like never before. And both their challenges and their successes have garnered the attention of the media and our national consciousness.

Recent articles in such publications as *Forbes*[1] and the *Harvard Business Review*[2] have chronicled the challenges facing women who enter the traditionally male-dominated fields of science, technology, engineering, and math (STEM). They must overcome cultural norms and expectations in schools, hostile or less-than-ideal workplace climates, and a lack of colleagues and role models. As Ellen Pollack posited in her 2013 *New York Times* article, "The most powerful determinant of whether a woman goes on in science might be whether anyone encourages her to go on."[3]

Fortunately, many initiatives today encourage women "to go on" in STEM. The National Science Foundation's ADVANCE program is dedicated to "increasing the participation of women in academic science and engineering careers."[4] National societies such as the Society of Women Engineers (SWE), Women in Science and Engineering (WISE), and the American Association of University Women (AAUW) are dedicated to supporting women's aspirations in STEM. A myriad of creative and engaging programs have sought to inspire and educate young girls: Invent It. Build It, Girls Who Code, EngineerGirl, MakerGirl, and SciGirls, and many more.[5]

Accounts of women in STEM are also proliferating. Some are compilations of biographical sketches: Anna Lewis portrays the lives of women architects, engineers, and landscape designers,[6] and Betty Reynolds and Jill Tietjen offer profiles of noteworthy women scientists and engineers through the ages, beginning with mathematician and astronomer Hypatia of Alexandria, born in CE 355.[7] Other authors examine women in STEM in the context of history. Margaret Rossiter[8] provides a detailed account of the societal forces influencing U.S. women scientists from World War II through the early 1970s. Martha Moore Trescott[9] documents the numerous women who became engineers between 1850 and 1980. And Margaret Layne presents the stories of successful women engineers and computer scientists[10] and employment trends[11] throughout the twentieth century in the United States. Amy Bix[12] focuses on the history of women in undergraduate engineering programs at Caltech, Georgia State, and MIT—capturing the challenges of navigating their education amid a climate where, early on, "the authority and tradition that had long made engineering a virtual male monopoly often made women feel like uninvited intruders in classrooms, laboratories, and residence halls" (p. 2).

Intruder is a strong but accurate word. Women in engineering have faced explicit and implicit biases, affecting their salaries, role expectations, self-efficacy, work environment, and other aspects of their lives.[13] And they've reacted accordingly. Trescott observed in 1990, "As women in this 'ultramasculine' field, engineers . . . have faced far greater hurdles than those which have had to be faced by either men or women who entered other professions. These women therefore had to be particularly persistent, determined, shrewd, and intelligent."[14] Bix in 2013 affirms that, despite past and present obstacles, "Today's young women take it for granted that they have the right to explore technical and mechanical interests, the right to enroll in even the most prestigious engineering and science schools."[15] So while these volumes provide important records of the lives of women engineers and the obstacles they faced, they also convey praise and hope for women in engineering.

Praise and hope is where this book comes in. The inspiration for *Women and Ideas in Engineering: Twelve Stories from Illinois* stemmed from our discovery of a volume from 1967 titled *Men and Ideas in Engineering—Twelve Histories from Illinois.* Published by the University of Illinois Press, it highlights "twelve accounts of men, events, and inventions in the hundred-year history of the University of Illinois' College of Engineering." *Men and Ideas* documents the contributions of individuals whose names and work are indeed part of the Illinois legacy: Arthur

Newell Talbot (engineering mechanics), Joseph Tykociner (sound on film), John Bardeen (the transistor and the theory of superconductivity), and William Fry (ultrasound), among others.

As the title indicates, all of these individuals chronicled in the book were men. In fact, the book mentions only one woman scientist: Elizabeth Kelly, a physicist and a research coordinator for the Office of Naval Research. In the 1950s, she advised William Fry from the Department of Electrical Engineering on his development of ultrasound equipment for studying the central nervous system. Apart from Elizabeth Kelly, women referred to in the book reflect traditional stereotypes: secretaries, housewives, and two cat-loving "matrons" asking Dean William Everitt to join the Humane Society (p. 96).

The use of the word *men* in the title, therefore, is accurate for its time. It was also the convention then to use *man*, *men*, and masculine pronouns as general terms representing both genders. Today, prose referring to a generic engineer or a researcher as *he*, or to Illinois's "pool of talented men" (p. 112) rankles us, as we value—and are accustomed to—more neutral and inclusive terms. In fact, our annoyance with the use of *men* in the title of the 1967 book prompted the writing of this one.

As we approached the 150th anniversary of the University, we realized that it was the perfect time to complement *Men and Ideas* by sharing stories of women engineers from our college, their impact, and their vision for their work.

Writing a title was straightforward enough; writing a whole book definitively less so. Our original intent was to write twelve stories that would be similar in structure to those in *Men and Ideas*. We therefore embarked on a mission to identify twelve women faculty members who had made significant contributions to engineering and to describe their accomplishments.

First, we visited the University Archives, where Bethany Anderson, Archival Operations and Reference Specialist, introduced us to their impressive collection. Bethany has been an invaluable resource and a delight to work with. She showed us original letters of recommendation for Rosalyn Yalow and local newspaper photos from the first campus meeting of the Society of Women Engineers. She also helped us over an unexpected hurdle: Early records of women students and faculty are sparse. We learned that, as essays by Chaudhuri et al. elucidate, women "were not until quite recently considered legitimate subjects of history and therefore of archival collection" (p. xiv).[16] In our case, for example, early graduation records did not include demographic data—so identifying 1917 Architecture graduate Marion Manley's gender took more hunting than we'd thought (photos in the campus yearbook revealed that Marion was indeed a woman). Nor has it been possible to accurately track other demographic data such as racial identity. We hope that our book leads to further collection of data and stories about underrepresented members of our College and of the field of engineering as a whole.

In addition to exploring the archives, we decided to use an abridged crowd-sourcing approach to help us. We asked more than two dozen faculty members, administrators, retirees, and alumni from the College to suggest names of women faculty who have made a significant impact; those women became our priority for inclusion in the book. These conversations also yielded so many fascinating stories about women graduates of the College of Engineering that we decided to include some of them as well—the first renovation of our original plan.

These early inquiries led to more remarkable and inspiring stories than we could have dreamed of. When at first we were uncertain whether we could find as many as twelve stories, we were later overwhelmed with possibilities. We decided to develop a companion website for the book so that we could chronicle more stories and add new ones as they unfold. This website is go.illinois.edu/WomenEngineers.

This abundance of stories led to the second significant revision of our original plan: Instead of featuring twelve women and their individual achievements, we would focus on twelve themes that reflect the significant impact of women from the College on the campus, in science and engineering and in society. This approach enabled us to highlight more than twelve women. While we primarily relied on the archives and our colleagues from the College to identify our highlighted women, we occasionally learned of a compelling story (such as Sakshi Srivastava's campaign to have a woman engineer statue on the engineering campus) that complemented a theme and demanded inclusion.

The feedback and encouragement that we received from our early sources thrilled us. It seemed our project was no longer "ours," but that it belonged to the College. Two graduate students in engineering, Sahid L. Rosado Lausell and Andrea Hupman Cadenbach, helped us gather preliminary data. Two other members of the college who were particularly engaged in contributing ideas and information were Celia Elliott from the Department of Physics and Cinda Heeren from the Department of Computer Science. We invited them, along with Bethany Anderson, to contribute further. Early on we met regularly as a group to plan and to discuss the women in the book, and our work together has been delightful. Another important milestone in our collaborative adventure was the addition of Amy Hassinger as our writing consultant. Her skills and insights have been a tremendous asset to our project.

When Cinda agreed to write a portion of chapter 7, "Women's Work: Computing," we realized that many other voices in the college could add unique perspectives. Hence the incorporation of first-person accounts, as well as short articles from colleagues in the college. Along the way, students like Katie Carroll, Sarah Rothe, and Catalina Hernandez helped. As a result, each chapter in our book is unique in structure and style.

We are grateful to everyone who shared stories of these remarkable women. Our interviews with many of the women featured, as well as their students, friends, and colleagues, have yielded not only rich insights and intriguing details, but also a deeper sense of connection for us as authors to these women and to the College of Engineering. Compared with *Men and Ideas*, our volume presents a more personal portrayal of many of the women we highlight. Our decision to use their first names aligns with that intention.

We have been exhilarated, intrigued, and challenged by these women's stories—some of which are being told for the first time. We have marveled at their creativity, determination, brilliance, and plain hard work. We have all benefited from their discoveries and inventions. We hope you will be inspired by them, too.

ACKNOWLEDGMENTS

We are indebted to Bethany G. Anderson, Celia M. Elliott, Amy Hassinger, and Cinda Heeren, who met regularly and cheerfully with us to bring this book to fruition and contributed significantly to the text.

We also thank others who wrote narratives that became part of the full story of *Women and Ideas in Engineering*: August Schiess, Steve McGaughey, Julia Stackler, Mike Koon, Jennifer T. Bernhard, Emily Scott, Claire Sturgeon, Katherine Carroll, Hui Lin Yang, Jonathan Damery, Ann-Perry Witmer, Christine des Garennes, Linda Reinhard, Laura Schmitt, Susan M. Larson, and Sakshi Srivastava.

We express our gratitude to Barry and Pauline Dempsey and to the College of Engineering for their continued support of this project.

Women and Ideas in Engineering

Engineers Who Happen to Be Women

The First 150 Years

ANGELA S. WOLTERS and BETHANY G. ANDERSON

The College of Engineering at the University of Illinois has its roots in the land-grant tradition. The passage of the Morrill Land-Grant Act of 1862 granted land to states for the establishment of colleges and universities dedicated to "branches of learning related to agriculture and the mechanic arts." Five years later, Illinois Industrial University emerged as one of the thirty-seven original land-grant institutions in the United States. The Morrill Act's stated purpose was to "promote the liberal and practical education of the industrial classes," a purpose Regent and first University President John Milton Gregory took to heart. Gregory envisioned an industrial education that included a rigorous engineering curriculum. Initially, this curriculum was offered through the Polytechnic Department, which comprised four schools: Mechanical Science and Art, Architecture and Fine Arts, Civil Engineering, and Mining and Metallurgy.[1] But by 1873, these schools organized under the new College of Engineering became the schools of Mechanical Engineering, Civil Engineering, Architecture, and Mining Engineering.[2]

Twelve years later in 1885, the Illinois Industrial University became the University of Illinois. At that time, the College of Engineering comprised sixteen faculty members and eighty-six students.[3] The College graduated its first woman student in 1879 but would not graduate another until 1897. During the first fifty years of the University, a total of eleven women would graduate from the College of Engineering. The number of alumnae from the College of

Engineering by the University's centennial in 1967 surpassed 100 with thousands more joining the engineering alumnae list by the school's sesquicentennial in 2017.

Nationally, many recognize the University of Illinois at Urbana-Champaign—the flagship institution of the State of Illinois—as a premier school for engineering education.[4] In 2016, the University of Illinois awarded the second most bachelor's degrees in engineering—2,111—of institutions in the country. Of those engineering alumni, women totaled 340, making them the seventh largest cohort of women receiving engineering degrees in the country. Also in 2016, the College's tenure/tenure-track faculty was comprised of seventy-six women, the third most of any institution in the country.

In the fall of 2017, the College of Engineering welcomed its largest class of freshmen women, 392, marking the representation of women at nearly 27 percent of the freshmen class.[5] Throughout the College of Engineering's history, the University of Illinois has been integral in the education of woman engineers.[6]

Those alumnae and female faculty have, in turn, influenced not only the University of Illinois but also the greater society. A recounting of the first 150 years sets the stage to celebrate the stories, experiences, and influences of many alumnae, faculty, and students who shaped and were shaped by the College of Engineering at the University of Illinois.

✳ Mary Louisa Page and the First Fifty Years: 1867 to 1917

Mary Louisa Page, the first woman student in the College of Engineering, began her studies in 1874. She was one of 89 female students on campus at the time, out of an undergraduate student body of 369.[7] The University admitted 23 women by 1870, which was "as early as suitable accommodations [could] be provided."[8] Within the College of Engineering, it was the field of architecture that drew the first women students, including Mary Louisa.

Born in 1849 in Metamora, Illinois, Mary Louisa attended Tremont Academy and Metamora High School before enrolling at the Illinois Industrial University. As a student, she was active in several groups and societies, including student government and the Alethanai Literary Society.[9] Established in 1871, the Alethanai Society—the women's literary society—provided Mary Louisa and the other members with opportunities to "better themselves in composition, elocution, debating powers and to enlarge their fund of general intelligence."[10]

Though Mary Louisa was one of several women on campus, she was the only woman in her architecture classes. Despite that challenge, she earned her bachelor of science in architecture in 1879, becoming not only the first woman

FIGURE 1.1 Mary Louisa Page, 1878. Photo courtesy of the University of Illinois Archives at Urbana Champaign, image 0006020.

to graduate from the College of Engineering[11] but also the first woman to earn a bachelor's degree in architecture in North America.[12]

She did have a few counterparts tackling similar situations at other institutions, including Margaret Hicks who graduated from Cornell University in 1880 and Anne Graham Rockfellow who graduated from Massachusetts Institute of Technology (MIT) in 1887.[13] At the time, women who decided to earn degrees and work outside the home were expected to work in "feminine" fields like nursing, education, or sewing. Women architects were pushing social norms.[14]

Following graduation, Mary Louisa established a drafting, blueprint, and abstracting service company—Whitman & Page—with fellow classmate Robert

Farwell Whitman in 1887.[15] Mary Louisa handled the drafting in the office while Robert served as surveyor and civil engineer. Choosing a male partner as her cofounder was undoubtedly strategic, a sound business practice for a female entrepreneur at the time. It probably helped put those customers uncomfortable with a female architect at ease. But Mary Louisa's role at Whitman & Page was on the forefront of a more widespread societal change. Women began to be seen as potential experts in architecture, particularly in areas such as residential design,[16] as early as 1901, when an article titled "Occupations of Women: What the Field of Architecture Offers to the Well Trained, Practical Woman" appeared in the *New York Daily Tribune*.[17]

Over the years, Mary Louisa held a variety of positions. She served as a teacher in Washington State, as President of the Women's Christian Temperance Union in Washington (1895–1900), and as an instructor at Blue Printing and Abstracting in Olympia, Washington (Alumni Register, 1913, Record Series 11/1/828). She wrote, too; her article "A Sketch from Life" appeared in the *Northwest Journal of Education*.[18]

By the 50th anniversary of the University in 1917, ten of the eleven women graduates of the College had been awarded bachelor's degrees in architecture. Louise J. "Prim" Pellens (BS Architectural Decoration, 1909) left behind a "Memory Book" from her student days. The book contains pictures of Louise at her drafting desk, programs from various campus events including the Architectural Club's Annual Banquet, and newspaper clippings about the social events of her sorority, Pi Beta Phi.[19]

While not a student of architecture, Nellie Nancy Hornor also studied in the College of Engineering during this time. Nellie received both her bachelor and master of arts degrees in physics in 1912 and 1913, respectively.[20] The Department of Physics had been established in 1889 as part of the College of Engineering, an organizational decision some deemed "unorthodox,"[21] but which allowed for greater breadth of research in the College and resulted in the additional involvement of women early in the college's history. Nellie was the only one of the eleven women outside the field of architecture to graduate from the College during the first fifty years of the University. Appendix A lists all eleven women.

Of the early architecture students, a few were highly notable contributors to the field, including Marion Manley. Marion, the only woman in her 1917 graduating class, chose to study architecture because she wanted to work in a field that allowed her to use both her brain and hands.

After graduation, she joined her brother in Miami, Florida,[22] where she found a job with Walter DeGarmo designing homes, many of them in the Spanish style.

FIGURE 1.2 Marion Manley at her drafting table, 1917. Photo courtesy of the University of Illinois Archives at Urbana Champaign, image 0003922.

A year later, she became the second woman licensed to practice architecture in Florida, and eight years after that she became the thirteenth female member of the American Institute of Architects (AIA). By that time, she had begun to design bigger projects, including Miami's U.S. Post Office and Federal Building.

While she designed over one hundred residences and commercial buildings over her sixty-year career, she is most recognized for her collaboration with Robert Law Weed on the master plan for the Coral Gables campus of the University of Miami as well as her design work on several campus buildings. Approximately forty years into her career, the AIA named her as the fourth woman fellow, a recognition given to those who have contributed significantly with a standard of excellence to the field of architecture, and she is referred to as "Miami's First Woman Architect."[23]

The real-world contributions of graduates like Mary Louisa Page and Marion Manley helped women see a place for themselves in the study of fields that were deemed "masculine." Over the next fifty years, more women from other subject areas began to join the College.

✳ The Next Fifty Years: 1918 to 1967

During this time period, the College of Engineering saw a rise in the numbers of women students, as well as in the milestones achieved by women students and faculty both, even as the college itself continued to evolve. New departments—such as ceramics, electrical, municipal and sanitary, and railway engineering—emerged in parallel with the technologies and professional exigencies of the time. And the College's departmental structure shifted as the world's needs changed. The Department of Railway Engineering, for example, had been created in 1906 to "provide training of a scientific character for those who wish to prepare themselves for the engineering, motive-power, traffic, or operating departments of either steam or electric railways."[24] But by 1940, having a department exclusively devoted to the subject was no longer necessary, so its courses were subsumed into Mechanical Engineering and other departments. In contrast, the Department of Electrical Engineering gradually expanded to include more laboratories and facilities.[25]

Together, the new fields of study within the College of Engineering allowed engineers and scientists to collaborate on interdisciplinary research and opened up opportunities for women. From 1918 to 1967, over 100 women would graduate with various degrees from the College of Engineering (detailed in Appendix B). Some followed in the footsteps of their predecessors by studying architecture and physics, while others pursued more traditional engineering disciplines. The milestones for women in the College came fast upon each other during this period of growth.

In the late 1910s, Beryl Love Bristow earned both a bachelor's (1918) and a master's degree (1919) in physics. Margaret Kate Dawson also earned a master's degree in physics (1919). Both Beryl and Margaret received their degrees at a time when the largest group of women to date—thirteen—were enrolled in undergraduate engineering study at the University.

Beryl was inspired to study physics by her own mother—the first woman to take a mathematics course at the University of Illinois and who then went on to earn a bachelor's degree in that same subject. Upon her own graduation, Beryl worked as a mathematician in the science department of Commonwealth Edison Company in Chicago. To recognize her pioneering spirit, Commonwealth Edison named a scholarship for women studying physics at the University of

Illinois in her honor.[26] In 1922, Eleanor Frances (Seiler) Wittman became the first woman to receive a doctorate in physics at Illinois. Several pioneering women in physics—including Lorella Jones, the first tenure track woman faculty member in the College (highlighted in chapter 2) and Rosalyn Yalow, the second woman to receive a PhD in physics (highlighted in chapter 5)—had significant impacts on their areas of study. Rosalyn eventually earned the ultimate recognition for her work as the winner of the 1977 Nobel Prize in Physiology or Medicine.[27]

The 1920s brought more milestones. Louise Woodroofe (highlighted in chapter 2), joined the College of Engineering as the first female faculty member. An artist by training, she taught architectural and freehand drawing. Grace Greenwood Spencer became the first woman to earn a degree from the University of Illinois in an "engineering discipline"—i.e., a specific field of engineering study where science is applied to transform abstract ideas into tangible products and technologies, all while designing under specific constraints.[28] Grace completed her chemical engineering degree in 1922, but did not graduate from the College, as the chemical engineering program was and continues to be housed in the College of Liberal Arts and Sciences. Instead, Carolyn Lindquist holds the honor of receiving the first official bachelor's degree in an engineering discipline from the College—fifty-six years after the establishment of the University. She graduated with a BS in Ceramic Engineering in 1923.

Nationally, women sought support networks of fellow women professionals. Two University of Colorado engineering students, Lou Alta Melton and Hilda (Counts) Edgecomb, tried to establish the American Society of Women Engineers and Architects (ASWEA) in the late 1910s and early 1920s. After sending letters to every college engineering and architecture department in North America, they were able to identify only 139 eligible women from 23 institutions. The fate of ASWEA had been correctly predicted by William G. Raymond, dean of the College of Applied Science at the State University of Iowa, who, in a reply to the ASWEA letter, stated, "You ask for information or suggestions. Have only this to say, that I suspect the number of women who have undertaken general engineering courses to be so few that you will hardly be able to form an organization."[29] Although ASWEA did not come to fruition from these early efforts, Hilda (Counts) Edgecomb did help to found the Society of Women Engineers nearly thirty years later.

One of the challenges in building supportive professional communities for women engineers and architects at the time was the national trend of moving architecture studies out of engineering programs, which left an even smaller support network. In 1931, the University of Illinois followed suit: it established the College of Fine and Applied Arts and moved the Department of Architecture under its aegis.[30] This change caused a sudden reduction of women in the College of Engineering, as the field of architecture had produced 18 out of the 25 female graduates from the College until this time. The last female architecture graduate to receive her degree from the College was Grace Wilson. Grace (highlighted in chapter 10) returned to the University in 1946 to teach engineering drawing and graphics. As the second woman faculty member in the College, she was a role model for and supporter of women students.[31]

When Mary Thelma Miller graduated with the first bachelor's degree in civil engineering in 1933, she was the only woman to graduate from the College with an engineering discipline degree that year and the eleven years that followed. World War II decreased enrollment during the late 1930s and early 1940s. However, the second half of the 1940s brought 16 female graduates from the College, among them the first female recipients of bachelors of science degrees in electrical and mechanical engineering—Lois Hume Windhorst and Anne Katherine (Lindberg) Williams, respectively—in 1945. Another prominent alumna with involvement in the early aerospace industry, Barbara "Bobbie" Crawford Johnson (highlighted in chapter 6) received the first bachelor's awarded to a woman in general engineering in 1946; that same year, Harriet Reese Wisely earned the first graduate degree in an engineering discipline, a master's of science in ceramics.

Additional "firsts" occurred in 1948 when Barbara Emily Jordan and Margaret Ellen (O'Donnell) Moran became the first women to secure bachelor's degrees in agricultural engineering and metallurgical engineering, respectively, while Ester W. Miller Tuttle became the first woman to earn a doctoral degree in an engineering discipline—ceramic engineering. Luckily for these aspiring women engineers, land-grant institutions such as the University of Illinois allowed women to enter the male-dominated field of engineering studies. Other technical institutions in North America—including Rensselaer Polytechnic Institute, Georgia Institute of Technology, and California Institute of Technology—were closed to women.[32] Throughout this time, "women studying or working in engineering were popularly perceived as oddities at best, outcasts at worst, defying traditional gender norms."[33] It should come as no surprise, then, that these pioneering women faced struggles on a daily basis. An article from the student-run *Daily Illini* entitled "Coeds Study Engineering—'Sole Girl in Class' Has Her Problems" from October 13, 1955, reveals that at the time only ten women were enrolled in the college. The article refers to these women as "fem-engineers":[34]

> One junior in engineering physics said that the first few weeks of classes were "awful" because she felt so self-conscious and there was no one to talk to. The freshman woman has a difficult time in classes the first few weeks. As work

progresses, however, interest picks up and she finds herself feeling more and more a part of the class instead of some kind of strange animal.

Not all women felt this way, of course. Some, like Betty Lou Bailey (highlighted in chapter 11), who in 1950 became the second woman to graduate with a mechanical engineering degree, had positive experiences. Betty Lou enjoyed her time at Illinois,[35] so much that she returned to campus as an alumna to counsel women students interested in engineering by sharing details from her own career with students and high school counselors.[36]

Still, enough women faced challenges that in 1959, a group of them gathered at the house of Professor Grace Wilson to discuss the formation of a student chapter of the Society of Women Engineers (SWE), a national organization founded in 1950. They faced their share of sexist naysayers, as evidenced by this article, which appeared in October 1959 in the *Daily Illini*, the student-run campus newspaper:[37]

> There is one place on campus where men outnumber women; so much so that the females are forming a protective society to improve their chances of succeeding in a predominantly male field. Seven of the 14 women enrolled in the College of Engineering have decided that there is safety in numbers no matter how small these numbers might be. These brave souls are attempting to create Illinois' first chapter of the Society of Women Engineers. What prompted the move was a feeling among the women that they could profit from each other by getting together and talking about their common engineering problems, . . . It is interesting that engineering females have problems different from men. Does it take a different sort of brain to use a slide rule, conduct experiments or study atoms?

Despite this kind of skepticism, the SWE student chapter officially formed in 1960. Four years later, they received their national charter. Since those early days, the SWE student group has flourished on the Illinois campus and currently serves as the largest registered student organization in the College of Engineering, with over four hundred active members.

As the 1950s drew to a close, only ten women were enrolled in the College, but that number grew slowly and steadily, reaching twenty-six by 1964. And nationally, sentiments about the inclusion of women in the engineering workforce were shifting, as statements like this one, made in 1963 by the Secretary of Labor, W. William Wirtz, attest:[38]

> It is clearly in the national interest to encourage qualified young women to enter the engineering field. The facts about women's suitability for an engineering vocation are so often interwoven with threads of fiction or prejudice that it is first necessary to separate fact from fiction. . . . Today there are approximately

7,500 women engineers—10 times the number in 1940; yet they represent only one percent of the profession. If more qualified women are to be attracted to this strategic occupation, encouragement and training must begin early—in high school. There is no lack of accredited engineering colleges which admit qualified women. Too often, however, young girls showing interest in and aptitude for scientific and technical subjects are not encouraged by parents and teachers. The failure to direct them to the prerequisite courses in high school forecloses their opportunity to enter the engineering curriculum in college.

Still, social norms are slow to change. Even though many people shared Secretary Wirtz's point of view about the necessity for harnessing the intelligence and work power of women engineers, still others habitually disparaged women who were stepping outside their traditional spheres. Witness the "Technocutie" column in the *Technograph*, the College of Engineering's student-run publication. The section was used to highlight a "cute" coed on campus. The March 1963 issue highlighted Miss Lois Backer, a general engineering student. The article ran with two pictures of Lois—a close-up and a full-length body pose (figures 1.3 and 1.4). It provides a taste of the kind of attitudes women engineering students regularly faced:[39]

MISS LOIS BACKER . . . WHAT? AN ENGINEER?

Increased rumors of several unidentified creatures—creatures resembling "girls"—have been circulating on the Engineering campus for some time. With the help of the Society of Women Engineers it didn't take TECH long (there were more than anticipated) to locate one, and here she is . . . lovely Miss Lois Backer—one of twenty-three charming women engineering students who are adding a touch of grace to our previously homogeneous engineering campus.

Lois was valedictorian of her Roanoke high school class, and she is now a 4.5 plus General Engineering freshman whose domestic talents acquired during five years of 4-H, should appeal to every engineer. Her beauty is self-radiating, but just for the records, she has been a Dolphin Queen semi-finalist as well as a Miss Woodford County contestant in the Miss America preliminaries.

Only one problem remains—where can TECH find another photographer? . . . Ours has found a new hobby—"Creature Watching"! As a matter of fact, where is the rest of the staff? Hey fellows . . . wait for me!

Nationally, though, the conversation sometimes acknowledged respectfully the special challenges that women engineers faced. In the summer of 1963, the national publication, *Engineer*, included a special feature titled "The Woman Engineer" that discussed the cultural climate for women engineers: "the women engineer deserves much credit for strength and courage. . . . Our entire cultural

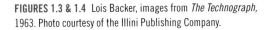
FIGURES 1.3 & 1.4 Lois Backer, images from *The Technograph*, 1963. Photo courtesy of the Illini Publishing Company.

atmosphere opposes women entering engineering and singles out this iconoclast as an 'oddball.' Our cultural stereotype maintains that a woman cannot be attractive, feminine and an engineer." It went on: "although engineering career opportunities have improved considerably since the postwar years, the lady engineer still comes across many industries whose welcome mats have no place for high heels."[40]

Such an attitude couldn't last long, though. Shortages in technical personnel meant that very few employers could continue to operate with such a stance. With time, women engineers saw opportunities open up in many industries.

By 1967, the university's centennial year, the College of Engineering could join the University as a whole in calling its past "distinguished," even as it looked forward to a "promising future": "From its traditional strengths in research and education to the promise of further gains through the application of new techniques and methods, the College measures its growth and plans its future in the areas of education, research, and public service."[41]

If the number and names of its departments are any measure, the college was most definitely evolving. By the close of the university's centennial year, the college comprised fifteen departments, many of them interdisciplinary: Aeronautical and Astronautical Engineering, Agricultural Engineering, Architectural Engineering, Ceramic Engineering, Chemical Engineering, Civil Engineering, Electrical Engineering, Industrial Engineering, Mechanical Engineering, Metallurgical Engineering, Mining Engineering, Nuclear Engineering, Physics, Sanitary Engineering, and Theoretical and Applied Mechanics.[42] At the centennial, twenty-eight women were enrolled in the College of Engineering—0.9 percent of total enrollment. The College had completed a significant phase of growth, which made it "ripe for some major changes."[43]

✳ The Following Fifty Years: 1968 to 2017

During this period, more female students began to enroll in the College of Engineering. By 1969, the percentage of women students finally reached 1 percent, and the upward trend continued throughout the 1970s, spurred on in part by "the influence of affirmative action, the women's movement, and high salaries." Nationally in the 1960s, women held 4.2 percent of positions as physicists, 26.4 percent of positions as mathematicians, and 26.6 percent of biological scientist positions.[44]

To assist in recruiting women to study engineering, the University of Illinois held a career counseling conference in 1973 titled, "Women in Engineering: It's Your Turn Now." Dean Daniel Drucker conveyed his hope that women engineering students would feel at home on campus: "A great virtue of a large university, like the University of Illinois at Urbana-Champaign, lies in the ability of a minority person to find enough others on campus to form a viable and strong group. You will be able to look around and see other women like yourselves and confirm that the choice you made was a sensible one. You will, in your academic work, be able to forget the fact that you are a woman. You will be an engineer."[45] Events such as this showed the commitment of the University to expanding opportunities for women engineering students.

Beyond career counseling programs, the College has traditionally held other events to showcase engineering as an exciting field of study for both men and women. One such event is Engineering Open House (EOH), which started in 1920 and became an annual event by 1952. EOH now brings tens of thousands of visitors to campus each spring.[46] In 1975, the EOH committee chose as the EOH theme "2001: An Engineering Odyssey." As part of the programming, a call was made to interest women in engineering studies. Two students, Carol Woodyard and Linda Aberle, wrote an article for the EOH program, "Women Learn about Engineering," in which they proposed the following:[47]

> Future Woman Engineer! Can you imagine a time when women will no longer be a minority of the engineers? The Society of Women Engineers' (SWE) Open House display has taken the view that in the year 2001 enrollments of women in engineering will be equal to that [sic] of the men.

Sadly, this vision did not become reality. During the late 1970s and early 1980s, efforts were nonetheless made to dispel the "myth" that engineering was solely a masculine profession. A brochure published by the Society of Women Engineers in 1979 wrote optimistically about increases in female enrollment in engineering programs: "[n]ot so long ago, an engineering school was fortunate to have 50 women students. Today, some schools enroll between 500 and 800 women students, bringing the total enrollment nationwide to more than 10,000."[48]

Despite these promising increases in enrollment, the population of women in the College of Engineering at Illinois stood at only 15 percent in 2001, not 50. Even fifteen years later, in the fall of 2017, only 22 percent of the undergraduate engineering class reports as female.[49] This trend parallels the national average, which shows somewhat stagnant growth of women studying engineering; numbers hover under 20 percent over the past three decades.[50]

This disparity in gender enrollment exists at the vast majority of engineering departments across the country. Over the years, engineering departments have made efforts to provide a supportive and inclusive environment for not only women but all students. In 1993, upon the recommendation from the Committee on the Status of Women Graduate Students and Faculty in the College of Engineering at the University of Illinois, the Women in Engineering (WIE) program was established. WIE's focus on the University of Illinois campus is to support the recruitment, retention, and personal development of women studying engineering disciplines.

The majority of programs providing dedicated support to women engineering students across the country, including WIE at Illinois, are members of the

Women Engineering ProActive Network (WEPAN), "a non-profit educational organization founded in 1990 to be a catalyst for change to enhance the success of women in the engineering professions." WEPAN is also dedicated to advancing cultures of inclusion and diversity in engineering higher education and workplaces.[51] WEPAN members work to champion these national efforts within their own local communities.

The efforts of the Women in Engineering (WIE) program at the University of Illinois are centered on providing effective programming to ensure equal access for both men and women, so both genders consider engineering as a potential career and both have equal educational experiences on campus upon deciding to study engineering. The WIE program offers engineering outreach to students of all ages, orientation camps for all incoming freshmen women engineering students, and customized professional development workshops including leadership training to provide personal skill development and career preparation for current students.

Specific retention efforts at Illinois include the WIE Freshman Orientation program, established by then Director Susan Larson in the fall of 2003. She created the program as a way for freshmen women to meet each other and build a community before fall courses began. Early formation of community is crucial to successfully navigating the transition from high school to college. In 2003, the University of Illinois' women engineering students were 15 percent of the incoming class.

WIE Freshman Orientation allows women engineers to move to campus early, become acquainted with one another and the campus, and form a supportive community of peers, upperclassmen, departmental advisors, administrators, and faculty members before classes begin. The program also teaches the students about the resource offices on campus and encourages students' enthusiasm and interest in engineering while addressing their concerns and challenges by creating a sustainable community of support.

Analysis of retention data of women students in engineering at Illinois has shown that those that attend WIE Orientations are more likely to stay, not only in the College of Engineering but also in their major of study. Efforts like WIE Orientation have led to an increase in graduation rates for women in the College of Engineering in the last decade. Since the initiation of WIE Orientation in 2003, the graduation rates of women in the College of Engineering increased over 17 percent. Students who attend WIE Freshman Orientation show a 30 percent increase in graduation rates as compared with those who do not attend. WIE Freshman Orientation has clearly helped improve the development of a pipeline of women engineering graduates to the workforce.

Women in the College of Engineering at Illinois have come a long way over the first 150 years of the University, but there is still much work to be done. We dedicate the stories that follow—stories of many notable women engineers with connections to the University of Illinois—to the goal of increasing general awareness and appreciation for the creative problem solving and significant contributions made by engineers of both genders. The women included in these pages have paved the way for others, advanced their fields of study, and contributed to the larger society. Their stories represent the work and impact of engineers, engineers who happen to be women.

Early Inspiration
Faculty Pioneers

LAURA D. HAHN

While Mary Louisa Page, the first woman student in the College of Engineering, started her studies in 1874, it would be another forty-five years before engineering students would encounter a female professor in the College—Louise Marie Woodroofe. She, along with two other pioneering women faculty members, Lorella Jones and Judith Liebman, is a treasured part of the legacy of the College of Engineering.

Their careers also reflect the developing nature of women's positions as faculty in science and engineering in the United States. In fact, neither Louise nor Lorella was an engineer, although their home departments (Architecture and Physics, respectively) were in the College of Engineering. Louise was an artist who became a faculty member at a time when women were considered "rare invaders" into engineering.[1] In Louise's era, drawing was a "prized traditional accomplishment for well-bred girls from Europe and America," and artistic skill was an early point of entry for a small number of women instructors in drawing for civil engineering and architecture.[2] Lorella Jones was a physicist—and during the decade in which she came to campus there were 1,577 doctorates awarded to women in the physical sciences, compared to just 77 in engineering.[3] It was not until 1972 that Illinois saw its first female faculty member who was an engineer: Judith Liebman. At the time, only about 1 percent of engineering faculty members were women nationally.[4]

✳ Louise Woodroofe (1892–1996)

The first woman faculty member in the College was an artist. After a childhood in Champaign and a few college years (1913 to 1917) at the University of Illinois, Louise transferred to Syracuse University, where she earned her bachelor's degree in painting in 1919. She then came back to Illinois in 1920 and joined the architecture department as an instructor of architectural and freehand drawing. She stayed for six years and then took a leave to focus on her own painting. When she rejoined the architecture department in 1930, she would spend only one more year in the College of Engineering, as architecture became a part of the College of Fine and Applied Arts in 1931.

Louise left a memorable impression. Those who knew her described her as "spectacular," "very opinionated, very talented," and "wild and splashy."[6] Her parish priest, Father George Remm, declared, "She was a very independent, strong-willed, and determined woman. She was a very creative and versatile artist. She was an excellent, innovative, and much-loved teacher. She was a fun-loving, humorous, needling friend." Borrowing, perhaps, from the French novelist Gustav Flaubert, who advised aspiring writers to be "regular and orderly in your life, so you can be violent and original in your work," Father Remm described Louise as the reverse: "orderly in her paintings," but chaotic in her life.[7]

> Another pioneer in the College of Engineering was Grace Wilson, who taught engineering drawing from 1946 to 1973. Her employment came about as a result of the manpower shortages from WWII. Although hiring women for such positions started as an emergency measure, it helped break down formal and informal barriers to women's participation in the engineering culture.[5] Grace's story is in chapter 10, "Mentors and Mentoring."

Townspeople recognized her by her cane—extravagantly decorated with colorful ribbons and flowers—and her orange-red bandana. And in fact, the theme of "color" is most prevalent in her colleagues' memories of her. Architecture Professor Kathryn Anthony wrote, "I remember on one occasion taking her to the TJ Maxx store and watching her carefully compare three or four red sweaters, taking great pains to select just the right shade of red. She wore crazy, colorful costume jewelry, often from the children's department." Professor Jack Baker wrote of "Louise in her blue smock and red bandanna giving of herself to the many students, helping them discover their hidden talents deep within."[8] Barbara Schaede, Louise's dear friend for many years, remembers her frequently asking to stop the car along a country road so she could stop to sketch purple and yellow flowers.

Louise's love of color shows through in her paintings. While Louise painted in a variety of styles and media, she is best known for her paintings of the circus. In the 1930s, she began spending her summers traveling with the Ringling

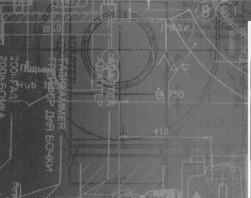

FIGURE 2.1 Louise Woodroofe, 1991. Photo courtesy of James Warfield, Professor Emeritus, School of Architecture, University of Illinois at Urbana-Champaign.

Bros. and Barnum & Bailey Circus, photographing and painting hundreds of circus scenes, including her specialty—clowns. And although her paintings were represented in prestigious galleries across the country, she was reluctant to show or sell them in Champaign-Urbana.

That is, until her students convinced her otherwise. In the 1950s, Louise was an adviser to the Illini Union Student Activities' fine arts committee. Those students pleaded constantly with her to show some of her paintings in the Illini Union. She finally agreed in 1960.[9]

Though there is no known documentation of Louise's experience as the only woman faculty member in the College of Engineering, it is clear that she had—and continues to have—an impact on students at Illinois. Max Slater, Architecture class of 1965 wrote in a letter of gratitude, "Miss Woodroofe led me to look at art and architecture and also most importantly myself and my goals in a better way. That has proved to be the most important thing I have learned in school."[10] Louise won the University's "Most Supportive Faculty Member" award in 1978. She established the Louise M. Woodroofe Award for a senior student in

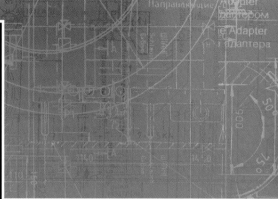

FIGURE 2.2 A drawing by Louise Woodroofe. Photo courtesy of James Warfield, Professor Emeritus, School of Architecture, University of Illinois at Urbana-Champaign.

architecture. And after her death in 1996, her former students, colleagues, and friends established the Louise M. Woodroofe Prize, which is awarded annually at the University of Illinois to recognize an architecture student based on his or her body of artistic work, including freehand drawings, watercolors, and architectural renderings and sketches.

Louise left her mark on the University of Illinois in another way: the configuration of the original Assembly Hall (now the State Farm Center). Its designer, Max Abramovitz, had been a student of hers. During construction, he invited her to take a look. Foreseeing a circus in the space, Louise recommended an entry door at the base that would be wide enough for circus animals and equipment and an apparatus on the ceiling for connecting a trapeze, features that Abramovitz incorporated into his design.[11]

For hundreds of years, art was an integral part of the design of devices—what we now call "engineering." While the fields diverged in the late 1900s, today's view of engineering once again embraces art and design. Louise's background as an artist and her role in teaching engineering drawing contributed an important element to Illinois's engineering curriculum—even today Illinois is one of the few engineering programs in the U.S. to offer freehand drawing.[12]

✳ Lorella Jones (1943–1995)

Lorella Jones was the first woman in the College of Engineering to achieve tenure. She came to the University of Illinois in 1968 as an assistant professor, and only two years later, at age 27, was promoted to associate professor. She became full professor in 1978. Lorella was a high-energy theoretical physicist at a time when few other women were, and she was an early adopter and promoter of Illinois's PLATO (later known as NovaNet), the first large-scale computer-assisted instructional system in the world.

Lorella was born in Toronto in 1943 to two scientists: her mother was an astronomer and her father was an industrial physicist. She earned a bachelor's degree in mathematics from Radcliffe College and a PhD in physics from the California Institute of Technology. Her research career in theoretical high-energy physics focused on applying the powerful tools of quantum field theory, which had been developed to describe the interactions of photons and electrons, to the multitude of new particles being discovered in the 1960s and 1970s. We now understand that these particles are *hadrons*, bound states of fundamental quarks and gluons. Lorella developed groundbreaking theoretical techniques to describe the evolution of these fundamental states of matter.

In the early 1980s, Lorella began exploring the use of computers to improve physics instruction in the department's large introductory courses. Efficiency was her motivation, according to her colleague Denny Kane. They developed computer-based quizzing to provide students with ample practice material and feedback while not encumbering the graduate students with the grind of grading. She became the director of the Computer-Based Education Research Lab (CERL) in 1992.

In an interview in January 2015, Ralph Simmons, former head of the physics department, observed Lorella's no-nonsense approach to work and life. "The thing that most impressed me," he said, "was her sense of herself, which was very straightforward. And she was as honest with herself as she was with other people. It was not self-criticism in any sense, just self-awareness. I think it was useful in her relationship with students, that honesty and straightforwardness." Indeed, her ruthless honesty, coupled with her insistence on academic rigor and high standards, characterize Lorella well in the eyes of her students and others who knew her. She was always keen to get things right: detecting flaws in arguments, challenging assumptions, and requiring evidence to make decisions.

S. P. Chia, a former graduate student, captured these traits in a letter of support for her promotion to full professor: "Her supervision was both critical and inspirational: critical because she would tear apart my calculations until she was satisfied with the result; inspirational because she attacked the problem at hand from different angles. I found her approach rather effective, as her criticism and inspiration drove me to work harder, especially at times of apparent despair." Similarly, Jeffrey Marque, a former graduate student who went on to work at Beckman Instruments, wrote to her in 1991: "Your courses were terrific, even if my performance in them was not. Surviving 480–481 with you did wonders to my sense of confidence. Since that bout with theory, and then an equally difficult bout (with a spectacularly successful conclusion) with an experiment as a post-doc in 1986–87, I'm not afraid of anything technical. . . . You were very good to me when I was your student." These comments portray Lorella as an exacting teacher, and she treated her undergraduate students with the same high expectations. Simmons described her 1979 undergraduate textbook, *An Introduction to Mathematical Models of Physics*,[13] as "a pretty rigorous book. It deals with lots of subjects that aren't ordinarily taught to undergraduates."

While Lorella's trademark was insistence on rigor and accuracy, Denny Kane noted that a "dry, ironic humor" often tempered her straightforward messages. One example is this note she wrote to her department head about what she saw as flawed departmental hiring practices and protocols for reporting work: "I am a bit put out by your letter mentioning prizes, etc., since it implies that those of us who are 'just doing our jobs' have somehow failed. Frankly, most of us think we deserve battle ribbons for keeping all the 'teaching, research, and service' balls in the air, with an oak leaf cluster for particularly disgusting jobs in 101 [introductory physics] courses."[14]

Another example comes from her classroom. Alan Craig, an undergraduate student in the early 1980s, recounts in an interview: "I had Professor Jones for Physics 106 [Mechanics, now Physics 211] my first semester at UIUC. She was great. I had it early in the morning (early for me!) so I always sat front and center to help myself stay awake and pay attention. I remember one day she was talking, and I was half out of it, and most of the other people were really out of it. While she was talking she subtly reached under the lab table, pulled out some kind of rifle, and fired it into a ballistic pendulum to demonstrate how you can measure the velocity of a projectile. That thing was LOUD! After she fired it she looked around the room and saw all the startled people. She looked at me and kind of smirked and said, 'Heh, heh, heh. . . . that gets 'em EVERY time . . .' Sometimes I think that shot must still be echoing around in Loomis Lab." (To be clear, firearms are no longer allowed on campus.)

On the more serious side, an essay that Lorella published in 1990 on "Intellectual Contributions of Women in Physics" exemplifies her predilection for setting the record straight; it also reflects her viewpoints on the challenges that

FIGURE 2.3 Lorella Jones playing the banjo. Photo courtesy of the Department of Physics at the University of Illinois and the Niels Bohr Library.

women in the sciences faced, beginning in the nineteenth century. In addition to providing an account of over a dozen female physicists and their scholarly contributions to physics research, Lorella describes how "women in physics have suffered their share of discrimination, disappointment, and trouble."[15] Early physicists such as Madame Marie Curie and Katharine Burr Blodgett attracted questions about whether they were "real scientists" and whether they would have been successful without scientist husbands to advocate for them. These women also had articles rejected when editors found out the authors were women. The essay goes on to express optimism about the decrease in discrimination by the 1990s. In contrast to the women in her essay, Lorella herself did not seem to experience such difficulties. Ralph Simmons summarized, "the fact that Lorella was a woman just never came up."

While Lorella had a no-nonsense nature and a weighty academic agenda, she was also known to have some fun—even outside of her mischievous responses to perceived breaches of appropriate academic conduct. She threw spectacular parties at CERL. She played the banjo, and she also enjoyed kayaking, gardening, and spending summers on an island in Lake Vermillion in northern Minnesota. Lorella prided herself on her Canadian heritage, although, as Ralph Simmons was quick to add, "She loved Illinois." She died from cancer on February 9, 1995, at the age of 51.

The department and the broader physics community mourned her loss, and today the physics department confers the Lorella M. Jones Summer Research Award on worthy students. In a 1995 cover letter to the editor of *Physics Today*, her departmental colleague Laura Greene described her as "phenomenal . . . truly extraordinary." She and her colleague wrote in their accompanying citation: "Professor Jones was a true pioneer. Through her writings and talks, she championed the cause of women in physics, and served as inspiration and role-model for the generations of women physicists who followed her."[16]

✷ Judith Liebman (1936–)

Judith Liebman was the second woman—and first woman engineer—to obtain tenure in the College of Engineering. She joined the Departments of Mechanical and Industrial Engineering and Civil Engineering as an assistant professor in 1972 and became a full professor in 1984. She was also a pioneer in administration: she began working as acting vice chancellor for Research and dean of the Graduate College in 1986 and officially took both positions in 1987. At the time, she was the highest-ranking woman ever to serve in campus administration.

Judith's interest in STEM began when she was a young girl. In an interview in December 2014, she said that when she was eleven or twelve, she read the book *Madame Curie: A Biography*, written by Curie's daughter Eve. The book was a gift from her mother, who continued to nurture Judith's interests. "Most of my Christmas presents and birthday presents were science kits. She let me have my own little chemistry lab in the basement, and I ruined one of her Revere stainless steel pans. I was doing an experiment with dyes and fabrics, and I had to buy her a new pan." Judith felt fortunate that her mother encouraged these interests.

Judith began her academic career with a bachelor's degree in physics but pursued a different career path. She said, "I enjoyed the mathematical modeling of things you could see, feel and touch. By the time I became a senior in physics, they were starting to worry about 'down in the atom and out in the stars'— things you couldn't see, and feel and touch." After she graduated in 1958, she returned to working with things you could see, feel, and touch: she took a job with Convair Astronautics to do calculations on the early Atlas missile flights. "I did that for about a month," she said, "and I realized that down the hallway, there was a computer. Between missile runs, we didn't have a lot of things to do. And so I told my boss, 'I'm going to go down and learn how to program a computer and then we don't have to do this on the calculator anymore.' So I ended up as a computer programmer." Here, Judith's initiative and inclination for efficient operations surfaced, and eventually these predilections guided her career path more explicitly.

She continued as a computer programmer, working for the General Electric Research Lab in Ithaca, New York, and then at Cornell. Her husband Jon, an environmental engineer, was minoring in operations research, a field of study that uses mathematical modeling to make decisions that optimize yield or performance. One day he suggested that she might like the field. Judith recounts, "I took an introductory course one semester, and I was hooked!" Again, the practical nature of the work excited her: "I was building mathematical models of things we could improve—as opposed to 'down in the atom and out in the stars.' So that's how I transitioned into operations research." She went on for her PhD in that field at Johns Hopkins.

After completing their PhDs, the Liebmans came to Illinois. Jon served in the Department of Civil Engineering, while Judith served both that department and Mechanical and Industrial Engineering. Judith said that her experience as the only woman in both departments was positive. "Fortunately, the industrial operations research group I was in had four younger male faculty members, and they just welcomed me with open arms." Gary Hogg, one of the members of that group, wasn't fazed by her gender, even though she was the first professional female he had ever worked with. Her talent and credentials impressed him. "It would be an insult to say that 'she fit in,'" he said, "it was far more natural than that—she belonged there. She was just one of us; working very hard to build a high quality program and graduate well educated students of which we would all be proud."

However, her gender did seem to catch others off guard, a reaction she found entertaining when she spoke about it years later. "About the first year I was teaching, the students walked in and looked surprised to see a woman," said Judith. "At the moment I was passing out copies of handouts. I said, 'The secretary will be here in a moment with more handouts.' The secretary walked in, and the secretary was a young man. And the look on the students' faces was absolutely priceless!" Later, after being on the faculty for a couple of years, the alumni board invited the new faculty to join their meeting in the Illini Union. As she walked over with fellow faculty members Dick DeVor and Gary Hogg, an older retired faculty member caught up with them and said to her, "You know, Judith, the wives are having the lunch on the second floor. Maybe you want to join the wives." Judith laughed, "I didn't even have a chance to respond. Dick and Gary picked up this guy, sat him on the sidewalk, and reamed him out good!"

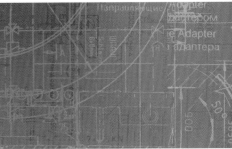

FIGURE 2.4 Judith Liebman faculty photo, ca. 1981. Photo courtesy of the University of Illinois Archives at Urbana Champaign.

These stories contrast with our perceptions of Lorella Jones, who seemed relatively immune to such incidents.

Both women did take an academic tack in addressing women's issues in their respective fields. In an echo to Lorella's essay on the status of women in physics, Judith published two articles on the status of women in engineering: "Women in Civil Engineering Education,"[17] and "Women in Engineering at the University of Illinois.[18] The latter article chronicled the increase in women students enrolled in engineering at Illinois from 1903 (0.1 percent, compared to men) to 1974 (5.7 percent). She attributed the increase to "energetic" recruitment activity from the college, as well as to societal trends for women to pursue careers. In this article Judith also pointed out that while more women were studying engineering, women faculty in engineering continued to be scarce. In a reflection of the challenges of her own work situation (and the academic climate of the day for PhD couples), she noted that women were expected to take research positions while their husbands followed the tenure track.

When Judith's husband became department head in Civil Engineering, she transferred fully to Mechanical and Industrial Engineering, where she achieved tenure in 1977 and became a full professor in 1984. Judith enjoyed research but was partial to teaching. And her partiality showed in her performance. One of her former PhD students, Alan Parkinson, is now dean of the Ira A. Fulton College of Engineering and Technology at Brigham Young University. In an email message in January 2015, he commented on her style: "She was energetic, very engaging, and used lots of examples. She passed out her notes for every session so you could easily follow along without having to frantically write down everything on the board. . . . I was attracted by the feeling you had in her class that it was OK to ask a question, or not understand something. She was very reassuring to students, that there were no 'dumb' questions or need for posturing. As a result, I felt motivated to really understand the subject." Judith won the College of Engineering's Everitt Award for Teaching Excellence in 1978 and 1986. She also became an Honorary Knight of St. Pat, the College of Engineering's Award for Service to Students, in 1981.

In addition to her teaching and research, Judith pursued numerous leadership opportunities, including serving on the Board of Directors of the Operations Research Society of America for eleven years. Gary Hogg remembers that her 1987 candidacy for president of that organization was met with skepticism because she was the first woman candidate. He wrote in a January 2015 email message: "While at our conference in a small room full of colleagues, a conversation broke out about how wrong it was for her to gain this position (largely because she was female was the implication). I shook my head, and said, 'Folks, this makes me ashamed to be in the same room' and I left." Despite that disheartening conversation, Judith won the election and became the first woman president of the board.

Judith also served on the Committee for Engineering Education of the National Academy of Engineering and the Advisory Committee for the National Science Foundation Engineering Directorate. At Illinois, she served as head of the industrial systems division in her department from 1978 to 1980. She also became involved in the faculty senate and assisted with departmental program evaluations and campus accreditation.

In about 1984, she organized a retreat on the use of computers in teaching, which got the attention of Chancellor Gerberding. He asked her to become the Acting Vice Chancellor for Research and Dean of the Graduate College; three years later, Chancellor Everhart appointed her permanently to the position. She approached her work with an operations research perspective: "After about five or six years, you see the same problems, but you can't solve them in the same way." Though she missed teaching, Judith enjoyed the decision-making process that accompanied her administrative appointments.

It is not surprising that Judith's enthusiasm and commitment to her work continue to have an impact. One of her former masters students, Terrie Reed, said in a phone interview in January 2015, "If it weren't for her, I wouldn't have the MS-IE (Masters of Science in Industrial Engineering) on my signature line." Terrie is now Senior Advisor for Unique Device Identification adoption with the Food and Drug Administration, and she remembers Judith as inspirational, "amazing," and "the gold standard." As Terrie's bachelor's degree was in social work, she often felt that "the male faculty in the department were waiting for [her] to fail." "Thanks to Judith," Terrie says, "I didn't. I think that was a victory for both of us, and it gave me confidence later on."

Judith's influence continues. In 2004, The Institute for Operations Research and the Management Sciences, the primary professional society of her discipline, established an award in her honor that "recognizes outstanding student volunteers who have been 'moving spirits' in their universities, their student chapters, and the Institute."[19] She inspired her colleague Gary Hogg to actively recruit women students and faculty in his positions at both Texas A&M and Arizona State. At both universities, Hogg recruited a record-setting number of female faculty, graduate students, and National Merit undergraduates. Hogg writes, "I attribute much of this to beginning my career with such a fine professional example as Judith S. Liebman." And her influence extends through the generations. In 2013, her granddaughter Lauren developed a PowerPoint report on her grandmother's career as an engineering professor. The report was featured on Rachel Levy's "Grandma Got STEM" website, a blog dedicated to celebrating senior women in STEM fields.

Today, Judith splits her time between Colorado and Illinois. She coteaches a class with her husband Jon in contemporary glass art at the Osher Lifelong Learning Institute at the University of Illinois. She has these words to say to women engineering students: "Get involved in an organization that involves leadership opportunities, like Engineering Open House. And take advantage of the breadth of campus."

Working in an era when women faculty were often marginalized and underutilized,[20] each of the pioneering women in this chapter approached their vocations with both acumen and individuality, influencing their students, their fields, and their communities. By the time Lorella's and Judith's careers were under way, women were joining the engineering faculty at Illinois at an increasing rate. Laura Eisenstein came to Physics in 1969, and Geneva Belford joined the Computer Science department in 1977. By 1985, there were eight tenure-track women faculty in the college. These small but important increases reflect national trends: Between 1977 and 1981, the percentage of women faculty in engineering, mathematics, computer science, and physical sciences increased significantly (from 10.2 percent in 1977 to 15.1 percent in 1981).[21] Because of developments such as the Civil Rights Act, the Equal Employment Opportunity Act, the National Organization for Women, and increased activism by the Society of Women Engineers,[22] "by the end of the 1970s, more progress had been made for women pursuing a career in engineering than in any previous decade."[23]

Research Orientations

LAURA D. HAHN

At its inception, the University of Illinois focused its resources on "dissemi-nating time-honored facts."[1] But in addition to excellence in teaching, the College of Engineering also valued research, and by the early 1900s, it began conducting testing associated with "boilers, locomotives, timber, and coal." Study expanded through the decades, so that by the 1920s and 1930s, the College supported "creative research dedicated to the development of new techniques and devices."[2]

Today, over 650 researchers in engineering at Illinois are involved in over 2000 projects,[3] including strategic research initiatives ranging from nanotech-nology to big data.[4] The faculty members in this chapter—experts in self-healing materials, flight control safety, radiation detection, and antennas—represent just a small segment of the many research orientations in the College. Jennifer Bernhard, Associate Dean for Research, describes her own research journey in a final reflection at the chapter's end.

✳ Nancy Sottos (1964–)

(Adapted from an essay by August Schiess and Steve McGaughey[5])

Collaboration, inspiration from nature, and "interesting science" are at the heart of Nancy Sottos's research. Working with a team of faculty members and graduate students, Nancy has been leading efforts to develop ways for composite materials (such as fiberglass in automobiles and airplanes) to heal themselves and therefore prolong their longevity and reliability.

As a graduate student, Nancy worked with a team in a center for composite materials at the University of Delaware. "It's not a surprise I wound up in a collaborative team," she says. "That's initially how I was trained to do research."

After earning her doctoral degree in mechanical engineering in 1991, Nancy joined the University of Illinois's Department of Theoretical and Applied Mechanics as an assistant professor. She would eventually serve as interim head of that department, from 2005 to 2006, as well as the chapter advisor for the Society of Women Engineers from 1997 to 2006. But research has always been at the forefront for her.

It wasn't long after she arrived at U of I that she started talking with Scott White from Aerospace Engineering about combining efforts to work with autonomous materials. They quickly recruited Jeff Moore from the departments of Materials Science and Chemistry, and together they created the Autonomous Materials Systems Group in 2001, one of the longest-standing research groups at the interdisciplinary Arnold O. and Mabel M. Beckman Institute for Advanced Science and Technology. According to Nancy, the Autonomous Materials Systems Group sits at the intersection of materials, chemistry, and mechanics: "We're focused on some central themes. We're inspired by biological systems and all their functions, in either plants or animals, and we're trying to reproduce these functions synthetically. We're trying to develop materials systems capable of self-healing, self-sensing, and self-cooling responses."

Take batteries, for example. "Batteries have significant reliability and safety issues," explained Nancy. "Nobody wants a heavy battery or one that's too large, especially in cell phones or electric cars. The way to make them smaller is to increase the density and capacity of the battery in order to store more energy. But as the energy density increases, they become unreliable and have greater potential for mechanical failures. A more serious issue is when they overheat, which is called *thermal runaway*."

Thermal runaway has become an infamous problem in cell phones and electric vehicles. The lithium-ion batteries that power these devices can overheat and catch fire, causing irreversible damage.

Nancy's research group studies materials that would autonomously shut down thermal runaway. One approach they've studied creates a microcapsule of polymers that are triggered to melt when a battery reaches a specific temperature. When the polymers melt, they block the flow of ions, and the process that is causing the battery to "run away thermally" shuts down.

FIGURE 3.1 Nancy Sottos. Photo courtesy of the Beckman Institute for Advanced Science and Technology/ Thompson-McClellan Photographers.

More recently, the group has been working on cooling concepts for the casing of a battery to keep the battery cool all the time, eliminating the need to shut the battery down. This technique would maintain optimal battery performance.

Nancy has also been working to solve the problem of battery capacity loss, which is especially problematic in electric cars. "If the battery capacity fades in a car, it can't go as many miles. So we're working on autonomous materials to influence the chemistry and the mechanical reliability throughout the life of the battery. Self-contained components will release additives or try to restore conductivity where it's been lost as the battery goes through its charge and discharge cycles."

Nancy is able to conduct this complex research in labs she shares with White and Moore. Moore's lab works with the chemical side of the research, such as synthesizing polymers and healing chemistries, while White's lab focuses on the processing of polymers and composites with the healing chemistries. Nancy's lab specializes in the characterization of the various functions of polymers with capsules and vascular networks, as well as molecularly changing the architecture of the polymer or composite to enhance the healing or cooling response.

It's been a team effort, from the work her lab has done to facilitate her research, to her long-lasting collaboration with White and Moore. Together they've contributed to the advancement of autonomous materials. Thu Doan is one of Nancy's PhD students who has been involved. She works on electro-spinning, which Thu describes as "a method where high voltage is used to turn solutions (liquids) into fibers (solids)." She goes on, "The fibers have a solid polymer shell that contains liquid material. When the material gets damaged, the shell breaks to release the liquid core material, which fills in cracks from the damage and hardens into a new solid material." Thu is using electrospinning to develop new protective coatings for steel.

Thu says that working with Nancy has taught her how to become an independent researcher: "Our group is large (somewhere around forty people), so we don't get a lot of one-on-one time with the advisor. So I've learned how to design and conduct my own experiments, as well as predict some of the questions that Professor Sottos might ask and design further experiments to answer those questions." Thu admires the high standards that Sottos sets. "She doesn't accept mediocre work. This raises the quality of work in our group. As student researchers, we know that our experiments need to be meticulously planned out and carefully conducted. I think this has helped me become a stronger researcher and pursue projects that are worthwhile."

One day, Nancy hopes to find the technology to not only repair but regenerate new synthetic materials. Regeneration includes both growing new material and remodeling old material. Working with her team, Nancy hopes to develop synthetic materials capable of this kind of regeneration in the next five to ten years. She also hopes to create more sustainable materials systems—materials that can both self-heal *and* self-report any damage, so that they might increase their life cycle. "My goals," Nancy adds, "are to maintain a great group of graduate students and continued success with my collaborators as long as we can. Not anything too lofty—just interesting science."

✳ Naira Hovakimyan (1966–)

(With Julia Stackler, Department of Mechanical Science and Engineering; Adapted from an essay by Mike Koon[6])

W. Grafton and Lillian B. Wilkins Professor of Mechanical Science and Engineering Naira Hovakimyan began her career as a mathematician educated in Armenia. Over the last nineteen years, researchers in the United States have used her research to help advance work in the stability of flight control systems. In March 2015, her research group's L1 control method was successfully tested on a Learjet at Edwards Air Force Base in California, and in 2016 the tests were repeated for the F16 Falcon Fighter Jet.

Naira, who is also a Schaller Faculty Scholar and a University Scholar, was a college student at the end of the Cold War. She completed a master of science degree in theoretical mechanics and applied mathematics in 1988 at Yerevan State University in Armenia, one of fifteen republics of the former Soviet Union. Three years later, Armenia seceded from the USSR. While the move brought freedom for Armenia, their fledgling financial system, along with an air and rail blockade from neighboring countries, crippled the economy and put most Armenians out of work.

"It was not clear how to live," Naira says. "We went five years without electricity or energy, living with oil lamps and candles. In those conditions, you don't wake up and think about going to work. People were exchanging real estate for a one-way ticket out of the country to go to a place where life was normal." Naira left for Moscow, where she completed a PhD in physics and mathematics in 1992 from the Institute of Applied Mathematics of the Russian Academy of Sciences.

Mathematics in Russia emphasized theory and beauty. "In Russian culture, as long as you impress another mathematician who may be more famous than you, then, 'wow,' and that's more or less the end of the story," Naira says. "In the Western world, people are looking for how it can be applied. In academia, if you want to get your students excited, you bring these applications into your group."

Naira knew she wanted to pursue applied mathematics in the academic world. But with conditions in Armenia in disarray, she would need to follow her interests in a different country.

From 1994 to 1998, she traveled the world, taking advantage of offers to study and teach. Because Germany was one of the first countries to establish an embassy in the newly formed Republic of Armenia, Naira applied for, and was granted, a scholarship as a German Academic Exchange Service scholar at Stuttgart University. She had similar opportunities in subsequent years in France and Israel, while also earning a Young Investigator Best Paper Award in Japan.

"It wasn't my original dream or plan, as I had no prior training in flight dynamics or flight control," Naira recalled. "However, a flight control system involves questions related to stability and robustness of performance and this problem is very mathematical in nature. Stability is something I had good training in back in Russia, so I took the challenge and joined the flight control group at Georgia Tech in 1998."

Naira spent the next five years in consecutive short-term appointments, with no guarantee of long-term employment. Absorbed in her research, she churned out impressive papers, which caught the eye of several sponsors of basic research, including the U.S. Air Force and NASA.

Bolstered by the funding, Naira landed a position as an associate professor at Virginia Tech in 2003. There, with the help of postdoctoral fellow Chengyu Cao (now on faculty at the University of Connecticut), she developed the L1 adaptive control theory, which can aid a pilot in regaining control of an airplane in sudden and drastic circumstances. Together, she and Cao wrote a book on the theory, titled *L1 Adaptive Control Theory: Guaranteed Robustness with Fast Adaptation (Advances in Design and Control).*[7]

The goal of L1 adaptive control in the aerospace industry is to help an aircraft maintain nominal handling qualities and minimize the challenge for the pilot when an aircraft fails (due, for example, to changes in aerodynamics, loss of control, coupling between control channels, or shifts in the center of gravity). If the airplane hits a wind gust and goes into stall (instability), losing lift, L1 adaptive control maintains the roll stability so the pilot can direct the nose down and recover.

"Once we developed it, NASA came along with opportunities of how to apply it," Naira says. "NASA was working on their own aviation safety program. They wanted to have an aircraft in a facility that could model flying in the wind tunnel." At the time, NASA was testing a 5.5 percent subscale general transport model aircraft in a variety of challenging conditions, including aggressive maneuvers.

"From the nine controllers that were attempted in flight, the L1 adaptive controller was the only one that could survive the stall and poststall conditions, giving the pilot a fully controllable aircraft. The L1 controller was eventually used for modeling unsteady aerodynamics in stall and poststall conditions, including the departure edges of the flight envelope," Naira says. Over the past twelve years, the system has been tested in a series of advanced conditions to prove that it works.

Naira accepted an invitation to join the faculty at the University of Illinois in 2008. Since then, her Advanced Controls Research Laboratory has maintained close ties to NASA. From 2009 to 2011, its Langley Research Center tested the L1 adaptive control system on the (unmanned) AirSTAR dynamically scaled Generic Transport Model research aircraft. The successful NASA Langley flight tests garnered Naira international recognition. She received the AIAA Mechanics and Control of Flight Award in 2011 and the prestigious Alexander von Humboldt Research Award in 2014. She was named an AIAA fellow in 2017. And in 2015, she won the Society of Women Engineers' Achievement Award—their highest honor.

In March 2015, the L1 adaptive control was successfully tested aboard a manned aircraft for the first time at Edwards Air Force Base. Over several weeks,

FIGURE 3.2 Naira Hovakimyan, 2016. Photo courtesy of Chad Olson.

students in the Air Force Test Pilot School performed rigorous evaluations on a Learjet in varying flight conditions. The team tested for seven failure configurations and the system allowed the aircraft to recover every time. Naira's technology was proven to be predictable, reliable, repeatable, and safe—the four response criteria that could set the stage for certification by the Federal Aviation Administration.

In September 2016, her L1 technology was tested on an F-16 at Edwards AFB. The tests were successful and confirmed once more her theoretical results.

Ultimately, Naira would like for the L1 system to have an impact on commercial aircraft. She noted that Boeing has invested "billions" in their flight control systems and likely won't replace them. "But, they could have ours as a back-up," she pointed out.

In addition to commercial aviation, Naira's team has had several inquiries from the drone market. The University of Illinois and NASA have patented the L1 adaptive controller with certain government rights.

Naira is also cofounder and Chief Scientist of IntelinAir, a company that brings actionable insights from aerial imagery to farmers to help them with timely alerts on problems in their farms. Naira has also consulted with Statoil, a Norwegian oil and gas company, to use the L1 adaptive control method in drilling applications. And Raymarine has commercialized the L1 adaptive controller for their Evolution autopilot for high-speed boats.

Naira's family is still in Armenia, where life has stabilized over the last two decades. She admits that she never anticipated staying in the United States this long, but she is proud of the fact that her hard work here is making a difference in air safety. "Illinois has given me a lot of opportunities and resources that have resulted in some great accomplishments. I am excited to see what we can continue to do in the future."

In addition to her extensive accomplishments in the area of adaptive control, in the last several years, Naira has also made headlines with her work on drone technology. She has been featured in the *New York Times* and on NPR, among other national media, for her "drones for the elderly"—which she anticipates becoming common among the rapidly growing aging population in the United States. These autonomous drones—still in development—will take care of everyday tasks such as bringing a person his or her medication, retrieving an item that's out of reach, and doing light household chores.

✳ Clair Sullivan (1975–)

"Figure out what matters," says Clair Sullivan, assistant professor in Nuclear, Plasma, and Radiological Engineering (NPRE).[8] Clair has applied these words of guidance to both her research and to her academic pathway. Her research involves using big data to identify radiation that could threaten national security—sorting consequential signals from irrelevant ones. Likewise, in her professional journey, Clair has sorted through her own experiences and ideas to elucidate what is and is not relevant for her.

Clair studied physics and astronomy at the University of Michigan as an undergraduate, participating in student research early on through a scholarship for women in science. Although she enjoyed the research, she resisted the idea of going on for a PhD. "And I will never forget the moment that I decided to go into engineering," she says. She had been invited to a small student gathering at a professor's house to meet Marty Pearl, a Nobel Laureate in physics. Someone asked him what his definition of intelligence was. Jokingly, Pearl described a continuum of the highest form of intelligence, with the top—science in its purest form—belonging to the mathematicians, while the "low of the low" were the engineers. "And that," says Clair, "is when I decided to become an engineer." She was determined to spite him.[9]

Clair then had some more sorting to do in her life. While she loved research, she was burned out on school; in fact, she had already accepted a job as a computer programmer. But her scholarship coordinator encouraged Clair to join the nuclear engineering department at Michigan. The plan was to get a one-year masters degree in order to get a good engineering job as soon as possible. But she quickly learned that if she wanted to have a senior position in industry, she would need a PhD. So she decided to stay on for her doctorate degree, conducting research with Michigan's premier group in radiation detector development. "I always really liked building things," she says.[10]

The September 11, 2001, terrorist attack prompted another revelation for Clair. She says, "I'd always believed that people should take whatever gifts they've been given and use them for the common good. So I decided that I wanted to go to Los Alamos to take what I learned during my PhD and apply it to counterterrorism."[11]

So after graduating with her PhD in 2002, Clair joined Los Alamos National Laboratory, where she worked with the Department of Homeland Security. Susan Mumm, Coordinator of Alumni Relations and Development for the Department of Nuclear, Plasma, and Radiological Engineering, describes Clair's achievements there: "Her work on detector development and deployment in the days after the September 11, 2001, terrorist attack earned her the lab's Distinguished Performance Award in 2004. With her contributions and technical and leadership skills, Sullivan quickly rose through the ranks to become the Senior Project Leader supporting the intelligence community. From 2008–2012, she worked for the federal government on nuclear and cyber-related matters. Her research and excellent communication skills also led to her being selected to brief members of Congress on Capitol Hill."[12]

After ten years at Los Alamos, Clair realized that it might be time for her to move on. She started exploring job postings for faculty members, and she joined the NPRE in 2012 as the first-ever woman faculty member in the department. "Every time I've said I don't want to do something, I wind up doing it," says Clair. "I never wanted to be a professor."[13]

Yet she says that coming to a Big Ten school "felt like home."[14] And she quickly found success at Illinois: "Clair's research at the interface of radiation detection, big data analytics, and new algorithms led to her selection as a winner of the 2014 Defense Advanced Research Projects Agency (DARPA) Young Faculty Award. The honor 'identifies and engages rising research stars in junior faculty positions at U.S. academic institutions and exposes them to Department

FIGURE 3.3 Clair with her daughter, Sophie Sullivan. Photo courtesy of Michael Sullivan.

of Defense needs as well as DARPA's approaches to developing emerging technologies.'"[15]

The project Clair worked on for DARPA centered on harvesting information through data analytics—sorting potential radiation threats from nonthreatening stimuli. An excerpt from a story on Clair by Susan Mumm describes this research:

> Current, commercial, off-the-shelf technology can be used to create very dense networks of radiation detectors. . . . The equipment, Clair says, "can pick up all kinds of nuisance alarms; it's a total needle in the haystack kind of problem." Her solution is to pull in other sources of data, including real-time weather, traffic patterns, geographic information system data, and other open-source data, to judge potential threats on the radiation network. "When you have a puzzle, what do you do?" she explains. "You start by grouping together like colors, then the edges, so you can start assembling."
>
> "Big data" is defined by the "four Vs": volume, velocity, veracity, and variety. The volume of data in true big data is considered to be larger than any single

computer or cluster of computers can analyze. Velocity refers to the fact that all data useful to a given project is in constant motion between the cloud and several series of computer servers. Data veracity suggests that not all of the data points within a data stream are correct, thus requiring complicated filtering mechanisms. Finally, proper analytics require the inclusion and analysis of a variety of data types to adjudicate an alarm [i.e., to determine whether a nuclear radiation alarm is actually valid, or is just a consequence of normal radiation fluctuation]. "Then you look for hits in the time-tagged, geo-tagged data," Clair says.[16]

Based on her work, Clair also received the 2015 American Nuclear Society (ANS) Mary Jane Oestmann Professional Women's Achievement Award. The ANS award cited her "for contributions in the areas of radiation detection, homeland security, nuclear nonproliferation, new course development, and for being an excellent teacher in the classroom."[17]

The course Clair developed on detector development is another reflection of her love of building things. This class is "modeled after a 'maker-lab' concept. The unique, hands-on class, first offered in the Spring of 2015 required that students use a Raspberry Pi (computer) to develop a radiation detector."[18] In this course, too, students have to figure out what matters: Clair encourages "extensive tinkering" and, in fact, gives a prize for the student who burns up her/his device first. Clair's love for teaching is clear to students: She received the department's Excellence in Undergraduate Teaching Award in 2013 and 2015.

Given Clair's energy and drive, it is not surprising that she pursues multiple interests outside of her academic life. She solders, runs marathons, swings on the trapeze, and skis—all a part of figuring out "what matters."

✳ Reflections by Jennifer T. Bernhard

Associate Dean for Research, College of Engineering Professor, Electrical and Computer Engineering

Men and Ideas in Engineering, the book that inspired this one, was published in 1967, a year after I was born. I own an original copy—I received it from one of my most influential mentors at the University of Illinois, Professor Paul Mayes. (His contributions, along with several of the other "greats" that established Illinois as a leader in antenna innovation, were highlighted in chapter 7 of *Men and Ideas—The Impossible Antenna*.) Paul gave me many of the books in his office when he retired for the "second" time—a transition from "part-time" to "full-time" emeritus professor—including *Men and Ideas*. I keep it on a bookshelf in my office as a reminder of how far we've come in nearly fifty years in the engineering profession, and how much farther we still have to go.

FIGURE 3.4 Jennifer Bernhard in 2004. Photo courtesy of the University of Illinois.

The title *Men and Ideas in Engineering* might seem exclusionary now, but it was simply a product of its time—near the end of an era, especially in academia, when engineering and engineering research was almost exclusively a pursuit of men. I was raised in the 1970s, the age of the Equal Rights Amendment movement, the first age in the history of our country when mothers and fathers, in large numbers, started telling their daughters that they could be anything they wanted to be. Perhaps a nurse—or perhaps a doctor. Perhaps a secretary—or perhaps a lawyer. Perhaps a science teacher—or perhaps an engineer. And, being a young impressionable girl like so many others, I took this parental guidance to heart. We have never let it go—and we never will. In the 1960s, our society underwent significant changes in regard to gender equality; it can take a while for societal changes to find their way into academia's cloistered environment. Now, of course, we know that each individual's contributions to innovation are not only important but also crucial—tackling broad societal challenges requires engineers with diverse sets of perspectives, ideas, and experiences. Indeed, the optimism and creativity at the heart of engineering necessarily lead us to this conclusion and are further evidenced by the collected experiences and accomplishments highlighted in these pages. I am optimistic about the continued growth of women in engineering, in part because of the inspiring stories captured in this book and in part because of my personal experience and collaborations with two of the faculty members profiled in *Men and Ideas in Engineering* so long ago.

My research focuses on the development of new classes of antennas and their associated design methodologies. My research team provides practicing engineers with knowledge and tools that can be applied in everything from mobile phones and sensors to radars and satellite communications. While Professor Mayes and his collaborators in the 1960s focused on the pursuit of *frequency independent antennas*, our research develops *reconfigurable antennas* that can respond as active components of the system to changes in operating requirements and conditions. When I began my career at Illinois, all of the faculty that studied antennas were either fully or partially retired. On paper, it looked as if I were alone—the first and only woman in electromagnetics and the only antenna specialist in the department—but luckily, Paul decided to welcome me into the Illinois antenna tradition. That is not to say that I didn't encounter bias and discrimination along the way—I did, and still do occasionally—but within the lab, I was in a sanctuary that let me pursue my passion and succeed. During the formative years of my faculty career, I collaborated with one of the greatest minds in our field and learned from him not only about different ways of thinking about antenna behavior and design, but also about the multiple dimensions of a faculty position, innovation, and even work-life balance. As my research portfolio grew, I was able to expand into interdisciplinary work with colleagues from other departments across the College, on wireless sensors embedded in concrete, electromagnetic systems for the regeneration of activated carbon cloth for environmental applications, and electromagnetic methods for tracking songbird migration from space. On this last topic, I was fortunate enough to collaborate with another world-renowned faculty member in remote sensing who was also highlighted in *Men and Ideas*, Professor Emeritus George Swenson. Both of these men of great intellect, skill, and character helped me fulfill my potential as a researcher, educator, and colleague, and I never had the sense that either of them thought twice about the fact that I was a woman. They treated me as an equal, and under their mentorship I thrived. I also credit these interdisciplinary experiences across the College for preparing me for the administrative position I now hold as Associate Dean for Research in the College, where my staff and I assist faculty in pursuing interdisciplinary research endeavors and the establishment of broad industry-based research centers.

The introduction of *Men and Ideas in Engineering* opens with the following statement: "Engineering may be defined in many ways, none of them altogether satisfactory. One of the reasons for this is that the field changes so rapidly." This statement has only become more true fifty years later and is evidenced by the breadth and depth of the research endeavor of our College. Engineering research continues to evolve and change, not only responding to current problems but also working to anticipate and prevent new ones. Even in just the last twenty years at Illinois, we have witnessed transformations and expansions of the definition of what it means to do research—including not only concentration on pure basic inquiry but also on more applied research; from almost exclusive government funding to the growth of industry-sponsored projects; from disciplinary concentration for the solution of specific technical problems to interdisciplinary team innovation that addresses large societal challenges. These new research models now lead to startup companies that address immediate needs and to fundamental advances that revolutionize their fields.

Some may say that increases in the numbers of women in engineering and engineering research in general is due to the emerging emphasis on tackling large societal challenges with interdisciplinary approaches. While that may be true, I also believe that women in our College have made these changes happen from within—the increased scope, the diversity of approaches, and the increasing adoption and success of interdisciplinary collaborative research are a result of the addition of women's perspectives, experiences, and engineering instincts in solving complex problems. We still have a way to go before we have equal representation of women in engineering, but we are headed in the right direction. The three accomplished individuals highlighted in this chapter are prime examples of thought leaders in engineering, and they represent a much larger set of contributions made to the research enterprise by women at all levels of our College over many years—and a foreshadowing of the engineering transformations that are bound to take place over the next fifty years.

4

Relentless Innovators

ANGELA S. WOLTERS

What makes one innovative? Authors of *The Innovator's DNA* identify five practices of the world's leading innovators: associating unexpected ideas with each other; asking "why," "why not," and "what if" questions; observing details and patterns in their world; experimenting—and failing; and networking with others.[1] Across time, these behaviors have led women to contribute numerous innovations to society: from the circular saw and windshield wipers to Kevlar and computer programming languages.

We also know that most innovations are not the product of a single person, but of a group of individuals. And the world is increasingly recognizing the value of diversity on teams. Katherine Phillips, dean at Columbia Business School, writes in *Scientific American:* "Diversity enhances creativity. It encourages the search for novel information and perspectives, leading to better decision making and problem solving. Diversity can improve the bottom line of companies and lead to unfettered discoveries and breakthrough innovations. Even simply being exposed to diversity can change the way you think."[2] Companies such as Intel, Apple, and Google are now investing millions in efforts to promote a diverse workplace.[3]

In 2016, the University of Illinois received the 2016 Higher Education Excellence in Diversity award from *Insight into Diversity* magazine.[4] While our efforts still have a way to go, they emphasize innovation with innovators working across disciplines to share technical knowledge. As of 2015, over 161 new start-ups associated with the University have been licensed and nearly 1,000 patents have been issued[5]—ranking the University of Illinois twenty-first in the world's list of the 100 top innovative schools.[6]

But innovation as it segues into entrepreneurship has proven to be a challenge, especially for women. Entrepreneurship is often noted as a "male phenomenon" because less than 10 percent of founders of high-growth firms are women.[7] Challenges persist in this field for women in the high-tech pipeline, given remaining stereotypes of women being less capable at leading new ventures, the lack of role models and mentors, and funding deficits—women-run businesses receive only 5 percent of venture capital (VC) funds. Other challenges such as life-work balance and fear of failure can also hold women back.

Many women from the University of Illinois have been innovators in their fields, working on teams and asking "Why not?" and "What if?" They have found various avenues of innovation in research and design (R&D) positions, academic pursuits leading to entrepreneurial endeavors and creative process development in industry. Some are highlighted in other chapters while the stories of three are shared here. Among them are Joan Mitchell, College of Engineering Hall of Fame Recipient in 2011, whose team at IBM invented the JPEG, and Jennifer Lewis, named one of 2015's Most Creative People in Business by *Fast Company*, who is advancing the use of engineered materials for 3-D printing. Students also innovate, as shown by the example of Amy Doroff, the 2015 Illinois Innovation Prize finalist. A commonality runs through all of these women's work—a constant, even relentless drive to find new, creative ways to solve engineering problems.

✳ Joan Mitchell (1947–2015)

With Celia M. Elliott, Department of Physics

If you've looked at or posted a picture on the internet, printed a color photograph, or sent or received a fax, you've used image compression technology invented by Physics alumna Joan L. Mitchell.

Joan was not the first female physicist in her family. Her paternal grandmother graduated from Stanford University with a bachelor's degree in physics in 1910 and earned a master's degree two years later. "I found out about her career when I was a junior in high school," Joan recounted. "My reaction was, 'If she could do it, I can do it!'"[8]

Joan described herself as a "Sputnik baby"—part of the generation of American school children that were pushed to study math and science after the United States was shocked by the Soviet Union's launch of the first artificial Earth satellite on October 4, 1957. "After Sputnik, we had to save the world from the Russians. It was okay for girls to be smart," Joan recounted.[9]

Joan's high school teachers recognized her talents for innovative science early when they selected her, one of seven hundred students, for the Bank of America's "outstanding science student" award. A National Merit Scholarship finalist, she enrolled at her grandmother's and father's alma mater: Stanford University. There Joan fell in love with physics. She worked on an independent research project to measure Brillouin scattering in liquid bromine, an experience that stimulated her interest in experimental condensed matter physics, which became the subject of her thesis research at Illinois. She was elected to Phi Beta Kappa at Stanford and graduated with a bachelor's degree in physics in 1969.

She came to the University of Illinois in September 1969 as a physics graduate student and received her master's degree in 1971 and her PhD in 1974. Prior to beginning her thesis research, Joan learned computer programming by working on the PLATO computer-based education project. Her new programming skills helped her with her thesis work, which involved the computer solution of a variety of extremely difficult nonlinear equations. She investigated and helped to elucidate a new physical effect involving the role of impurities in diffusion in insulating materials.

Immediately after receiving her PhD from Illinois, Joan went to work for IBM Corporation in the Exploratory Printing Technologies group at the T. J. Watson Research Center. In the course of her thirty-five-year career at IBM, she proved to be one of its most talented and prolific inventors. The holder of more than 70 patents, Joan was one of the top 20 inventors in IBM's research division. Her work was crucial to the development of coders and decoders used in data compression standards for facsimile machines, printers, and digital transmission of images.

Joan made major contributions to IBM's development of the Group III facsimile standard that was adopted by the Comité Consultatif International Téléphonique et Télégraphique (CCITT), the organization that sets international standards for data communications. She successfully lobbied for implementations that were as suitable for software as they were for hardware, which led to a simpler product.

Joan subsequently popularized a modification of the Group III standard within IBM for computerized storage (versus facsimile transmission) of binary images. Many IBM products have successfully used this technique, from teleconferencing equipment to high-speed printers for mainframes.

From 1987 to 1994, Joan was a leader in the Joint Photographic Experts Group, the organization that developed JPEG image compression, the first color still-image data compression international standard. Crucial to modern

FIGURE 4.1 Joan Mitchell (left) and Gerald Goetzel (right), recipients of IBM's Outstanding Innovation Award, 1979. Department of Physics, University of Illinois at Urbana-Champaign, courtesy AIP Emilio Segrè Visual Archives.

communications, the JPEG standard consists of 20 explicitly defined processes to encode or decode continuous-tone still images. Joan headed the U.S. delegation in 1991–1992 and also served as an editor of the standard's final documentation—a crucial role in ensuring its dissemination and acceptance.

At IBM, Joan made major contributions to *arithmetic coding* technology for data compression, which can smoothly adapt to changing event probabilities, making it superior for documents containing both text and images. Arithmetic coding was included as part of the JPEG (still-image) compression standard, and it is a key component of the newer JBIG binary-image compression standard. Joan led development of a reduced-computational arithmetic coder called the *Q-coder*, which replaced the multiplications in arithmetic coding with additions.[10]

Joan Mitchell was elected to the National Academy of Engineering in 2004 "for leadership in setting standards for the formation of photographic fax and image compression." In 2011, she received the Institute of Electrical and Electronics Engineer's Masaru Ibuka Consumer Electronics Award "for fundamental contribution to image compression in printing technology and digital imaging processing in consumer electronics." She was also an IBM Fellow and a fellow of the IEEE.[11]

Joan has referred to her work as "a series of lucky coincidences." Those who knew her well, including her coauthor of *JPEG: Still Image Data Compression Standard*, William Pennebaker, shared that it was her "intelligence, persistence and hard work" that led to her innovative work and accomplishments.[12] Joan was so persistent, in fact, that it sometimes interfered with her sleep: "I can't sleep for the good reasons," she reported in a 2003 interview. "I'm so busy inventing that I can be up 10 or 12 times a night. And maybe about six o'clock I'll finally sleep until noon. So I get the sleep, but meanwhile I've had a pretty exciting night."[13] Saturday, she noted, was her "fun day" because without the phones ringing she could accomplish a week's worth of research in a day. Only church on Sundays and her scuba diving classes—she was a dive master—kept her from working through weekends.

Throughout her career at IBM, Joan remained connected to the University of Illinois. In 1996, during a sabbatical from IBM, she served as a visiting professor in the Department of Electrical and Computer Engineering and visiting scientist at Beckman Institute for Advanced Science and Technology. In 2005, Joan returned to campus for a week to conduct a series of mentoring visits for women engineers and undergraduate physics students.[14] During that visit, she shared insights about mentoring, work-life balance, problem-solving, and networking: "I was well into my IBM career as a scientist before recognizing the importance of written goals, remembering names, networking, giving back, reaching back, selling your ideas, and leaving an auditable paper trail."[15]

A couple of years after her visit, Joan went on to publish a book. *Dr. Joan's Mentoring Book: Straight Talk about Taking Charge of Your Career* grew out of a personal project: Joan wrote daily emails to those interested in hearing her stories—mentees, former students, fellow engineers—that illustrated important lessons on mentorship and career advancement. As both a mentor and an inventor of new technologies, Joan made a positive, innovative impact on the world.

✳ Jennifer Lewis (1964–)

Jennifer Lewis (BS Ceramic Engineering, 1986) began her studies at the University of Illinois in the fall of 1982. She applied to the College of Engineering as a general engineering student but was recruited to ceramic engineering on a merit scholarship. Although Jennifer had some exposure to engineering—her father worked as an engineer for General Electric—she knew little about materials science, but she followed her interests. She is now the Hansjorg Wyss Professor of Biologically Inspired Engineering at Harvard's John A. Paulson School of Engineering and Applied Sciences,[16] having joined the faculty in 2013 after a twenty-year career at the University of Illinois.

Throughout her life, Jennifer said in a 2015 interview, "I have repeatedly benefited from being a bit naive and simply taking 'leaps' as opportunities arose."[17] Those brave leaps have resulted in notable advancements in the field of materials science, such as her invention of new classes of printable material and 3-D printing techniques.

Jennifer had several important role models while growing up in the northwest suburbs of Chicago, her parents and several high school teachers among them, but one teacher stands out: Mrs. Kolder, her high school English teacher. Jennifer's high school years established a strong foundation of knowledge for her to build on as an undergraduate student at Illinois.[18]

For Jennifer, the choice to pursue Ceramic Engineering at Illinois was the right one—"It was a small, cohesive department and I had the good fortune to have several outstanding faculty and older student mentors," including Professors David Payne, James Nelson, and Trudy Kriven. She particularly appreciated the way Professor Nelson served as an informal undergraduate advisor.[19]

As an undergraduate, Jennifer thrived in and out of the classroom, building skills that have contributed to her success. An avid sports enthusiast, Jennifer participated in numerous campus intramural team sports. She served as president of Keramos, the materials science honor society, and also as a resident advisor in her dormitory. These experiences provided Jennifer with an understanding of the importance of the team (over the individual) and developed her ability to lead by building a strong team. Throughout her career, she has built interdisciplinary teams that feel empowered to innovate together.

Ultimately, Jennifer decided to pursue her graduate work in ceramic science at Massachusetts Institute of Technology (MIT). The unusual properties of ceramics interested Jennifer: "They can form glass, porcelain, and clay but also readily conduct electricity and are a key material in many high-tech electronics." She loved how varying production and assembly could affect the materials' properties. Ceramics processing, she felt, was "part science, part art."[20]

After earning her doctorate degree, Jennifer—with a passion for research and teaching—decided to launch her own research program at Illinois. She returned to campus "with the hope of having a positive impact on the next generation of students."[21] And this is just what she's done.

FIGURE 4.2 Jennifer Lewis. Photo courtesy of UI News Bureau.

Analisa Russo, cofounder of Electroninks Incorporated, shared that working as a graduate student with Jennifer was a "magic combination."[22] Analisa was interested in the maker community and had taken a class in the MIT media lab before enrolling in her PhD program in Materials Science and Engineering at the University of Illinois. At MIT, Analisa learned "How to Make Almost Anything" through a course titled as such.[23] It inspired her to think creatively about how to use tools like 3-D printing. Analisa married her passions for materials and STEM education in her graduate PhD work with Jennifer and fellow student, S. Brett Walker, in the creation of Electroninks, a company focused on commercializing conductive inks for the printed electronics and do-it-yourself (DIY)/maker community.

Over the past twenty-five years, Jennifer's research interests have grown. Initially, she focused strongly on colloidal processing of ceramics, including their phase behavior, rheology, and colloidal assembly. After receiving tenure in 1997, Jennifer took her first sabbatical and started a fruitful collaboration with a colleague, Dr. Joseph Cesarano, at Sandia National Laboratories in Albuquerque, New Mexico. "This opportunity opened up an entirely new research direction, 3D printing of ceramics, for my group," she shared. "Since that time, our research has vastly expanded to include the development of new classes of printable materials, microfluidic printheads, and 3D printing platforms."[24]

With the goal of designing and fabricating "matter that matters," Jennifer and her research group have moved ideas from the lab to the commercial sector. She finds it "gratifying to see some of the ideas . . . have a chance to have a major impact."[25] Witness Jennifer's research in bioprinting, which earned her recognition as an innovator from *Foreign Policy* as one of the 100 Leading Global Thinkers of 2014.[26] Or her role as CEO and founder of Voxel8, a company that has created the first 3-D printer capable of producing fully functional electronics—a position that led *Fast Company* to name her one of 2015's Most Creative People in Business.[27]

Jennifer's distinctions don't stop there. A member of the National Academy of Engineering, Jennifer won the National Science Foundation Presidential Faculty Fellow Award, the Brunauer Award from the American Ceramic Society, the Langmuir Lecture Award from the American Chemical Society, and the Materials Research Society Medal, and she is a fellow with the American Ceramic Society, the American Physical Society, the Material Research Society, the American Academy of Arts and Sciences, and the National Academy of Inventors.[28]

But the accolade she's most proud of is the Material Research Society (MRS) Medal. Established in 1990, the MRS Medal recognizes exceptional achievement in materials research over a period of ten years. Jennifer is the first woman to receive this award. According to Jennifer, "it really reflects the contributions of my research group, my students, undergraduates, graduate students, postdocs, and our many collaborators. . . . I just get to be front and center."[29]

Jennifer—author of over 160 papers and holder of twelve patents—believes that "innovation is an evolutionary process—one that is not done in a vacuum. Our innovations emanate from fruitful interactions with members of my research team as well as our collaborators." She's built an interdisciplinary research team that continues to push forward the practice of 3-D printing:[30]

> At the heart of my work is the simple notion that 3D printing can do more than create complex shapes (or forms). We have therefore strived to integrate form *and* function into 3D printed devices by developing new materials, printheads and printing methods. Consequently, our research has led to many breakthroughs, including high performance electrically conductive inks, polymer inks, colloidal inks, filled epoxy inks, elastomeric inks, and, most recently, cell-laden inks for both planar and 3D printing. We have leveraged this broad palette of materials to achieve first demonstrations of 3D printing of omnidirectional silver microelectrodes, electrically small antennas, capacitive fiber sensors, 3D polymer scaffolds with submicron resolution, and lightweight structures based on epoxy and elastomeric lattices.

Looking ahead, Jennifer is most excited about her emerging work on 3-D bioprinting of vascularized tissues. "We are addressing major societal needs in drug safety screening, tissue repair and regeneration. Importantly, this work began less than four years ago—hence, it is still in its infancy. However, we have already made important foundational advances that are propelling the entire field."[31]

When asked about the future and what she still hopes to accomplish, Jennifer's first response is, "I have so much more to learn and do!"[32]

✱ Amy Doroff (1992–)

Adapted from an essay by Emily Scott[33]

As an intern at John Deere in the summer of 2014, Amy Doroff (BS General Engineering, 2015) worked on improving the process of installing lock collars during combine assembly.[34] She led a team to design a brand new process and tool that decreased warranty claims, improved safety, reduced ergonomics issues, and allowed for international process standardization, work that won her recognition as an up-and-coming innovator. But she almost didn't get there.

Six weeks into Amy's freshman year in General Engineering at Illinois, she wanted to quit. "I didn't know what I wanted to do, I just couldn't do physics," she said. "Engineering wasn't easy. The equations spinning in my head had me questioning whether or not I was really cut out for it. I wasn't used to struggling in classes, so I showed up in the department advising office tired and ready to throw in the towel. Thankfully, I ran into an advisor who wouldn't let me quit. I stayed in General Engineering and I'm glad I did."

That advisor was Heidi Craddock, assistant director of undergraduate programs and chief academic advisor for the Industrial and Systems Enterprise (ISE) department. "Amy's such an analytical thinker, problem solver," Heidi said. "You can just see it any time you talk to her, the wheels are always turning. . . . I could see it in her that she had what it took, she just needed someone to believe in her and remind her to believe in herself."

Amy used the support from Heidi to set her sights on completing her degree and doing it well. Over the next few years, she focused on being, as she puts it, relentless. She and fellow engineering student Savannah Goodman (BS Civil and Environmental Engineering, 2013) would often ask each other, "Have you been relentless today?"

"I liked that word," Amy said. "I was relentless in getting involved in my residence hall, including mentoring students as a resident advisor. I was relentless in attending office hours 5 days a week as a freshman to close the gap between my classmates and me. That word guided me to work hard to prepare for every exam and it led me to look for ways to test my engineering skills." In the end, Amy's relentlessness paid off. She began thriving in her course work and extracurricular involvement.

One day during her junior year, Amy noticed a sidewalk recruiting tent for John Deere. She spoke with a recruiter and landed an internship. "I like to say I got a Frisbee and a job," she said. "More or less, that's what happened. . . . I had secured a summer internship that would lead me to opportunities I had never thought possible."

That summer, she went to work at John Deere's Harvester Works factory in Moline, Illinois. Two weeks into her internship, she was assigned to a high-priority engineering project that involved working on a lock collar. This part was found on rotating shafts within combines. Its purpose was to stop the shaft from moving laterally by locking it to the bearing, but the process for installing the lock collar was unreliable, making it ergonomically unsound for assembly-line operators. John Deere had also received associated warranty claims.

Over the next twelve weeks, Amy buried herself under a pile of lock collars while she struggled to find a solution. One night, while cooking a risotto after a long day of work, the answer came to her: "I mostly went hungry that night, but my idea had set off a snowball of positive consequences. The next morning I came into work with the sunrise. I went straight to the toolmaker and asked him to build a prototype for me." Over the remainder of her internship, she examined nearly forty-five locations in the factory that linked to her work on the lock collar, often working late to test her ideas. She gave demonstrations, talked to bosses and operators on the manufacturing floor, and centered her efforts on this one tiny piece of a combine.

By the end of her twelve-week internship, Amy had created a tool that fixed the ergonomic and warranty-claim issues associated with the lock collar. Amy said she thinks she was able to make such an effective, award-winning tool because she could see the problem from multiple angles. In high school, she spent four summers assembling patient food trays at Advocate Health Care, a job that gave her a perspective that she believes most engineers don't have. "When I went to work at Deere, I had the ability to come up with things in different ways because I knew it from the operator side as well," she said. "The main solution I made for a project was to make it safer to come to work every day. And it happened to be something that could be globally implemented and standardized across the enterprise."

Harvester Works leadership took notice of what she and her team had created and nominated the project for the Dan Levengood Excellence in Ergonomics Award at John Deere's Global Ergonomics Conference in September of 2014. The day she attended the conference, still ignorant of the award she was about to win, Amy looked around the room at her fellow innovators from Deere, some of them three decades her senior. "I thought, maybe I'll be as good as them one day," she said. "And maybe I'll have the opportunity to come back and actually have a shot at this."

As the only intern and the youngest person presenting with a team, she never considered she would have a shot at winning the top prize. The Moline factory had never won the award. With no expectations, standing in front of representatives from all over the world, she spoke from the heart: "I relayed information about just how much warranty claims would be lowered. I relayed stories of operators who would go home with their hands and shoulders intact and I got excited talking about automated validation processes to ensure repeatable processes every time."

She watched a few presentations, then left early—she wanted to get back so she could go to her Control Systems class the next day. Later that evening, she got a call informing her that her project had won the award. "I was speechless," she shared. "I still couldn't figure out what to say when I called my parents 10 minutes later so I just repeated 'we won.' An hour later I was still in shock." The Dan Levengood Excellence in Ergonomics award is the top award globally in the ergonomics category for John Deere.

Back at school, Craddock nominated Amy and her John Deere project for the Technology Entrepreneur Center's Illinois Innovation Prize, an award that recognizes the most innovative student on the Illinois campus. "When we saw the call for nominations for that prize come through, she immediately came to mind, because we had talked about her project, what she had done, and why she had implemented it," Craddock said.

When Amy got the email informing her she had been nominated, she couldn't think why. "I thought I had some strong accomplishments, but I still wasn't in that headspace of thinking of myself as one of the best innovators on campus," she said. She submitted an application and made it to the interview round, where for the second time in a year she found herself in a room surrounded by people she thought to be far better innovators. "I was just struck, thinking: 'I hope to be as good as them, I hope to be worthy of being in this room one day,'" Amy said. She ended up making it to the final round as the only undergraduate student to do so. She was awarded the third place prize and $1,500.

"The judges were quite kind to me," Amy said. "They said they recognized a lot of potential in what I had done, and I had been the only undergraduate in the competition, and my project had also been over the span of less than twelve weeks—some of the PhD research that was going on had been over the span of several years."

After the award ceremony, she was approached by a representative from a company asking if she'd assist them with some tool design. So she decided to

FIGURE 4.3 Dean Andreas Cangellaris congratulates Amy Doroff, 2015. Photo courtesy of William Gillespie.

use her award money to start her own company. She and Alex Rubocki, a fellow GE senior and Amy's lab partner from her Control Systems class, incorporated Hunter Madison Design LLC, a small tool design company that is named after two of Amy's golden retrievers. "Neither of us really wanted our name on the company, but my dogs didn't care."

In the summer of 2015, Amy worked as an assembly engineer for Whirlpool Corporation at their Amana, Iowa, location. Before she started her job, she wrote the word *Whirlpool* on a piece of paper and wrote her job description goals. "Then I said, but what are my people goals? And what are my goals to sustain the organization as a whole? And what are my goals in other categories?" she said. She began to find opportunities outside of her job description that could add value to the company, including being a recruiter, building the relationship between Whirlpool and the ISE department, and working on a morale initiative. "I think that really speaks to the ISE curriculum. The ISE background has taught me that a true leader is not just successful at meeting the goals in their job description."

Above all, she had enthusiasm for everything she could do at Whirlpool, including manufacturing and operations, which has always been her passion.

"Manufacturing works well for me, because I'm a morning person," Amy said. "Manufacturing can start at six or seven o'clock in the morning, so often I'd drive to work with my headlights on. And I'd get there, and I'm yelling, 'It's a great day to make refrigerators!'"

Her love for manufacturing led her to be accepted into Whirlpool's three-year manufacturing leadership development rotational program that began in January of 2016. Now you might find Amy anywhere from Tennessee to Iowa, making things like refrigerators, ranges, and small appliances. "I have a lot of enthusiasm for making things. I started in operations," she said. "I used to be the assembler and I used to try to come up with new innovative ways to get the trays down the line faster. Now, as an engineer, I just have a broader chance to do that. I'm really excited to go to Whirlpool and have a shot at that. And hopefully one day, I'll make my way into different leadership areas once I've learned the whole company." After learning all parts of whatever company she works for, it would be Amy's dream to have a position such as a factory manager or regional director.

The fact that Amy so readily contributes much of her success to the help she has received along the way is what Craddock believes makes her a great leader. "There are not that many leaders that are willing to do that," Craddock said. "But those that do will go a lot farther, and people will want to follow them. Amy will draw people to her because they want to be a part of that."

Amy once thought she couldn't do physics. Now, thanks to her own relentlessness, she's been a physics tutor, a successful graduate, and an acknowledged innovator, preparing for her future as a leader in her field.

For students like Amy interested in the innovative entrepreneurial ecosystem, the College of Engineering launched a new Innovation, Leadership, and Engineering Entrepreneurship (ILEE) dual-degree program in the spring of 2017. The program allows engineering students "to better understand the innovative processes involved in identifying complex technical problems and creating, developing, and leading efforts to provide for their engineering solutions."[35]

According to Andreas Cangellaris, Dean of the College of Engineering, together the Faculty Entrepreneurial Fellows—established in 2015—"gives faculty financial resources and time away from their classroom and committee obligations to pursue commercialization of their innovations"[36] along with the ILEE program, "fundamentally redefines the role of faculty in innovation"[37] and allows students to "invest their time and energy in entrepreneurship and innovation as they study engineering."[38] These programs enhance the opportunities for engineering faculty and students to be relentless in innovating the world.

Acclaim in Bioengineering and Medicine

LAURA D. HAHN

Nationally, more women today are going into biomedical engineering than all other engineering disciplines except for environmental engineering. In 2016, 41.4 percent of bachelor degrees in biomedical engineering went to women.[1] This relatively young field may attract women because it lacks the "guys only" tradition of more established engineering fields. In addition, as many observers have pointed out, "Women seem to be drawn to engineering projects that attempt to achieve societal good."[2]

The societal good pursued by the two women in this chapter is human health. Rosalyn Sussman Yalow earned her PhD in physics from the University of Illinois in 1945. Her work on radioimmunoassay (a technique for detecting hormone and insulin levels in the blood) earned her the Nobel Prize in 1977. Princess Imoukhuede, a faculty member in the Department of Bioengineering, is breaking ground developing nanosensors for cancer research and continuing Rosalyn Yalow's legacy of advocating for women in STEM.

✳ Rosalyn Yalow (1921–2011)

Celia M. Elliott, Department of Physics

September 18, 1941. While Europe is engulfed in war, twenty-year-old graduate teaching assistant Rosalyn Sussman attends her first meeting of the faculty of the College of Engineering. A few months earlier, Rosalyn boarded a westbound train in the South Bronx, where her family lives. She came to Champaign, Illinois, to study physics at "the most prestigious school to which [she] had applied."[3] She is the only woman at the meeting of 400, and the Dean congratulates her.

October 13, 1977, 6:45 a.m. Thirty-six years later, Rosalyn Sussman Yalow is at work in her office at the Veterans Administration Hospital in the Bronx when the telephone rings.[4] She has won the Nobel Prize in Medicine or Physiology, the first American-born woman to win a science Nobel and the second woman to win a Nobel Prize in Medicine.

The daughter of immigrants who never attended high school, Rosalyn decided by the time she was eight years old that she would be a scientist. By seventh grade, she had settled on mathematics, although she later chose physics.[5] About her career in science, Rosalyn modestly said, "I wandered into it, oddly enough, because of things I couldn't do. I couldn't draw. I'm tone-deaf, so that rules out music. I've got no athletic talents. My chemistry teacher taught me the joy of working with my hands and my brain."[6]

She graduated from high school at age fifteen in 1937, and her parents advised her to be an elementary school teacher. "That's what bright Jewish girls did in the thirties," she explained.[7] But she was determined to be a scientist. There was no money for college tuition, but her grades won her a coveted spot at Hunter College, the highly competitive women's college that was part of the New York City Colleges, where tuition was free to residents.

Deeply inspired by Eve Curie's biography of her mother, published in 1938, Rosalyn eagerly chose physics as her major. "In the late thirties when I was in college, physics and in particular nuclear physics, was the most exciting field in the world."[8] While Rosalyn was excited about achieving a career in physics, it at first seemed like an impossible dream—most good graduate schools were highly unlikely to accept and offer financial support to a woman. The professor at Hunter who recommended her for admission to Purdue University received the following reply: "She is from New York. She is Jewish. She is a woman. If you can guarantee her a job afterward, we'll give her an assistantship." No job guarantee was possible during the depths of the Depression.[9]

With medical school an impossibility and graduate school unlikely, one of Rosalyn's Hunter physics professors, Jerrold Zacharias, arranged a part-time job for her as secretary to Rudolf Schoenheimer, a biochemist at Columbia University's College of Physicians and Surgeons, thinking that she could sit in on graduate classes that way. Rosalyn could type, but she had to agree to learn shorthand to get the job, so after she graduated from Hunter in January 1941, she went to secretarial school. "Fortunately I did not stay there too long. In mid-February, I received an offer of a teaching assistantship in physics at the University of Illinois. . . . It was an achievement beyond belief." The $70/ month teaching-assistant's salary and free tuition meant that she could tear up her stenography books and take the next step toward fulfilling her childhood ambition.

Despite her Phi Beta Kappa key and excellent grades at Hunter, Rosalyn discovered that she had taken fewer physics courses than the other first-year graduate students at Illinois. She threw herself into her work, auditing two undergraduate physics classes and enrolling in three graduate classes her first semester, in addition to teaching half-time. She got all A's except for one A– in the optics lab.[10] Years later she still smarted from then-acting department head Gerald Almy's comment, "That A– confirms that women do not do well at laboratory work."[11] She never forgot or forgave it.

She also met her husband, Aaron Yalow, on her first day of graduate school and it was, as she later described it, "Interest at first sight."[12] Because of University nepotism rules (husbands and wives could not both be employees of the University) at that time, the two did not marry until 1943, when Rosalyn received a fellowship and thus was no longer a teaching-assistant employee. She described her time in graduate school thus: "The campus was filled with young Army and Navy students sent [there] by their respective Services for training. There was a heavy teaching load, graduate courses, an experimental thesis requiring long hours in the laboratory, marriage in 1943, wartime housekeeping with its shortages and rationing, and in January 1945 a Ph.D. in Nuclear Physics."[13] As part of her thesis research, and exploiting her prodigious skills in mathematics and chemistry, Rosalyn became adept at building instruments to manipulate and measure radioactive substances during her time at Illinois.

As soon as Rosalyn received her PhD in nuclear physics in 1945, she returned to New York City to look for work. She was unable to find a position with a research university, so she became the first woman engineer[14] at the Federal Telecommunications Laboratory, the research division of International Telephone & Telegraph Corporation. When the research group she was working in left New York in 1946, she returned to Hunter College to teach physics. However, Hunter lacked the laboratory facilities for Yalow to continue in research, and her career as a nuclear scientist was going nowhere.

In the meantime, Aaron Yalow had accepted a position in medical physics at Montefiore Hospital in the Bronx. Through him, Rosalyn Yalow met Dr. Edith Quimby, a leading medical physicist at the Columbia University College of Physicians and Surgeons. Rosalyn volunteered to work in her laboratory to gain experience in the medical applications of radioisotopes, which were just then being produced in sufficient quantities at Oak Ridge, Tennessee, to make them readily available and cost-effective for therapy, diagnosis, and biomedical investigation.[15] Quimby in turn introduced Rosalyn to Dr. Gioacchino Failla, a pioneer in biophysics and radiobiology and "The Chief" of America's medical physicists.

As Rosalyn reported it, "After talking to me for a while, [Dr. Failla] picked up the phone, dialed, and I heard him say, 'Bernie, if you want to set up a radioisotope service, I have someone here you must hire.' Dr. Bernard Roswit, Chief of the Radiotherapy Service at the Bronx Veterans Administration Hospital, and I appeared to have no choice; Dr. Failla had spoken."[16]

Rosalyn joined the Bronx VA as a part-time consultant in December 1947, building and calibrating radiation detection equipment and planning experiments for the safe use of these materials in medical diagnosis and treatment. Her engineering experience was invaluable; commercial instrumentation was not yet available, and Rosalyn designed and built most of her own equipment. It soon became apparent to her that the most important applications of radioisotopes would be in elucidating human physiology and as an aid to clinical diagnosis.[17]

Still teaching at Hunter College full time, Rosalyn started to build a full-fledged Radioisotope Service at the Bronx VA, starting from a tiny laboratory in a former janitor's closet. In January 1950, Rosalyn decided to leave teaching and joined the VA Radioisotope Service full time. She soon became acquainted with Dr. Solomon A. Berson, a young physician who was just completing his residency in internal medicine at the Bronx VA. In July 1950, Berson joined the Radioisotope Service and began a twenty-two-year research partnership with Rosalyn that would culminate in the Nobel Prize for the discovery of radioimmunoassay (RIA). Theirs was an ideal team—Rosalyn had the experience in physics, radiochemistry, and mathematical modeling, and Berson had the clinical expertise—to make huge strides at the interface of physics and medicine.

The RIA technique that Rosalyn and Berson invented allowed scientists to measure precisely minute concentrations (one-billionth of a gram) of antigens in the human body, such as hormones in blood. Contemporary reports compared the achievement to the ability to measure a teaspoon of sugar dissolved in Lake Erie.[18] In addition to its exquisite sensitivity, RIA had a number of other significant advantages. The assay could be performed in test tubes—no radiation entered a patient's body—and tiny amounts were sufficient for the tests. Before RIA, a diabetic had to give 100 cc of blood (roughly one cup) for each blood test; RIA could be performed on 0.1 cc.[19] Finally, RIA could be used to detect virtually every human hormone and important biological molecule and could be performed in any laboratory equipped to measure radioactivity.

Rosalyn and Berson published their idea for RIA in 1956,[20] but their concept was slow to be adopted—it was too new, too complicated, and too unbelievably sensitive to win converts in the conservative medical establishment. Rosalyn reported that when Berson gave a speech at the University of Illinois, "One person in the audience thought he would win a Nobel Prize and the other 29 thought he was crazy."[21] But by 1970, RIA had been widely adopted, and it revolutionized endocrinology. In addition to measuring human hormones, today RIA is used in blood bank screening for the hepatitis virus, early cancer detection, the diagnosis and treatment of peptic ulcers, and research in neurotransmitters in the human brain. Thousands of children have been saved from certain mental retardation caused by congenital thyroid defects through newborn diagnoses made possible by RIA.

Rosalyn spent much of her scientific life as an outsider—first as a graduate student in physics at the University of Illinois and then at the Bronx VA, where she was a woman PhD in a new field in a hospital dominated by male physicians and military officers.

Her former student, Mildred Dresselhaus, now a chaired professor of physics at Massachusetts Institute of Technology, said of Rosalyn, "The only way she could make her point was to be very precise, definitive, and assertive. Sometimes people see her brusque side. But to be noticed in the world of science, she had to be that way. She was an outsider in every way. She was working in a new field of physics, and she didn't have the right credentials in medicine. So she had to let them know she was for real."[22]

Rosalyn herself attributed her success to four personal qualities: "I've always been well organized. I've always considered what I wanted. I've been prepared to work for it. And I've given much thought to the complications associated with the paths I had set myself upon."[23]

For many years, Rosalyn kept a creased and faded photograph in her desk.[24] It showed her as a young child in a New York park, wearing man-sized boxing gloves and looming over her older brother Alexander, sprawled at her feet as if she had knocked him out. Her parents had staged the photo to commemorate Rosalyn's triumph at school that day. Five years earlier, Alexander's first-grade teacher had smacked his hand with a ruler for some infraction, causing him to burst into tears and throw up, much to his mortification. Five years later, when the same teacher cracked Rosalyn's knuckles, instead of crying, Rosalyn grabbed the ruler and hit back.[25] Marched to the principal's office and told to account for her behavior, Rosalyn explained that she had been waiting for years to avenge her brother. "That's the attitude that made it possible for me to go into physics."[26]

FIGURE 5.1 Rosalyn with her medal, in the clothes she wore for the Nobel ceremony. Photo courtesy of the Department of Physics at the University of Illinois and the Niels Bohr Library.

At Columbia University's first Saturday Science Symposium for young people in 1982, a child asked Rosalyn about the discrimination that she had faced as a woman and as a Jew. She replied, "Those who are discriminated against feel hostile and second class, so it's important to develop an inner sense of security. Courage comes from within you."[27] Although she was opposed to sexual discrimination, Rosalyn did not identify with feminism. "I don't join women's groups. I prefer to be where the power is, and that's with the men." She added, "There was never a time when I didn't do what I wanted. . . . I don't want anyone telling me I can't do something because I'm a woman."[28]

Rosalyn elaborated on her thoughts about the discrimination that she had faced in her professional career in a speech at the Nobel Banquet on December 10, 1977. Traditionally, the students of Stockholm present a tribute to the Laureates, and one Laureate is chosen to respond. That year it was Rosalyn. She began her remarks: "The choice of one among the several deemed truly and equally distinguished must indeed be difficult. Perhaps I have been selected for this privilege because there is certainly one way in which I am distinguishable from the others. This difference permits me to address myself to a very special problem."[29]

The problem was, she said, the underrepresentation of women in the Western world as scientists and leaders. "No objective testing has revealed such substantial differences in talent as to account for this discrepancy. The failure of women to have reached positions of leadership has been due in large part to social and professional discrimination. . . . Even now women with exceptional qualities for leadership sense from their parents, teachers and peers that they must be harder working, accomplish more and yet are less likely to receive appropriate rewards than are men. These are real problems which may never disappear or, at best, will change very slowly."[30]

Rosalyn is one of only two women in the College of Engineering Hall of Fame (along with Joan Mitchell—chapter 4, "Relentless Innovators"). Rosalyn's words to the students of Stockholm should inspire all of us: "If women are to start moving towards that goal [of equal representation in science], we must believe in ourselves or no one else will believe in us; we must match our aspirations with the competence, courage and determination to succeed; and we must feel a personal responsibility to ease the path for those who come afterwards. The world cannot afford the loss of the talents of half its people if we are to solve the many problems that beset us."[31]

✳ Princess Imoukhuede (1980–)

(Adapted from an essay by Claire Sturgeon[32])

Almost forty years later, Princess Imoukhuede is answering Rosalyn's call. As an assistant professor in the Department of Bioengineering, she is bringing her passion and expertise to solving the problem of cancer. Concerned that more than twenty American women are diagnosed with cancer every hour, she says, "The situation inspires revolutionary approaches toward improving patient treatment and survival."

Most types of tumors, including cancer, require a supply of blood to grow larger than a few millimeters. Scientists have made great progress in combating cancer by finding effective ways to stop angiogenesis, or the formation of new blood vessels. Princess and her colleagues have furthered that progress by engineering ways of personalizing angiogenesis-inhibition cancer treatments.

Princess's lab uses nanosensors to better understand the tumor microenvironment and why the same type of tumor may behave differently in different people, just as two mulberry trees might react differently to the same herbicide.

"My lab is trying to understand whether there is a subset of patients for whom anti-angiogenic treatments are especially useful, and if so, find out how we identify those patients," said Princess, who is also an affiliate of the Carl R. Woese Institute for Genomic Biology. "That's where we get into the area of personalized medicine, being able to tailor anti-angiogenic treatments specifically to a patient."

The cells that make up tumors have different populations of receptors that promote blood vessel growth. Princess says these receptors can serve as biomarkers, helping doctors predict drug responsiveness by providing a quantitative way to profile cells.

"If there is a traffic jam and you block a freeway, you'll find that cars will go through some of the side streets. We can try to block some of those side streets, but cars will still try to find a way through," Princess said, describing the way anti-angiogenic drugs block receptors that encourage tumor growth. "This is the problem with cancer research, where you block one marker, receptor, or molecule, the tumor still finds another way."

For personalized cancer treatments to become a reality, Princess says scientists must understand the tumor microenvironment, find a way to count the number of receptors, apply that data to computational models that predict cancer drug efficacy, and suggest the best treatment options for each patient.

They're working on it.

"The most exciting take-home message is that we are able to find certain cells within the tumor microenvironment that we haven't profiled previously," Princess said. "We determined that a certain subset of these cells had very high levels of expression of one of these angiogenic receptors that could actually negate some of the effects of a common anti-angiogenic drug."

That anti-angiogenic cancer drug, Avastin, already has been developed and is approved for many types of cancer, including brain, lung, and colorectal cancer. However, the Food and Drug Administration revoked the drug's approval for metastatic breast cancer due to evidence that the survival benefits did not outweigh the side effects for many patients.

Princess's research may someday make these drugs available to a subset of metastatic breast cancer patients who might benefit more than others from the survival benefits of Avastin.

Princess credits her family with planting "engineering seeds" early on that led to her commitment to biomedical research:

> My brother is a law professor now, but back in the olden days, he'd come home from kindergarten and teach me how to count. He'd come home from class and teach me how to add. He'd come back from children's choir and teach me how

FIGURE 5.2 Princess (L) in her lab with student Grace Conard, 2013. Photo courtesy of UI News Bureau.

to sing. Whatever he was learning, I had to learn too! Now, if you consider the 2.5-year age difference, you can imagine that oftentimes his teachings would be met with blank stares—but he kept trying to teach me and I kept trying to learn. I think that he helped me learn how to process things quickly. He also helped make learning fun. He planted seeds that learning was important.

My dad is a civil engineer—he studied structures and materials at the University of Illinois in Chicago. As far back as I could remember, he would sit me down to do math problems. When I wasn't doing math problems, I was often running off with his triangular architectural ruler—using my markers to draw straight lines with it. I, of course, needed these straight lines to draw houses along with their indoor layouts. Little did I know, I was absolutely ruining his ruler! He'd replace it, and I'd run off with the replacement. I'm sure it was a terrible cycle for him. So, I would say that my dad's patient encouragement of me in my very important drawings and his math teachings were some important seeds.

My mom was another influence. As far back as I could remember, she was at a university campus, taking classes. So until I was about four years old I attended daycare at Governors State University. She also took classes at Chicago State University and Prairie State College. Some of my early memories are running through campus halls, watching the "cool" college kids buy whatever they wanted from the vending machines, making funny faces to them through their classroom windows as they learned, and watching them study in the campus cafeteria. I became very comfortable in universities—they became like a second home to me. So, I'd say my mom's love of learning, her determination to earn her degrees, and the importance of learning were additional seeds that were planted.

My mom eventually worked as a teacher and administrator, which I'd say certainly inspired my current profession. She is also an outstanding cook! She taught me how to bake cakes, cookies, muffins, biscuits, and traditional Nigerian meals. I loved trying out different recipes, playing around with temperature, and seeing what would happen when you added too much of this or less of that. I learned very quickly that too much baking soda does not a good cookie make! One of my favorite memories is making cookies on the BBQ grill with my best friend, Valerie. I think these baking experiments helped me explore the same kinds of questions scientists and engineers explore—what happens if I change this parameter or if I change the temperature or the instrument. Again . . . more engineering seeds.

At a point, my parents became concerned that my love of baking might limit my ideas of what was possible—so they transitioned my baking interests towards traditional scientific inquiry by buying me a children's chemistry set. I kept trying to find ways to make those inert chemicals explode—of course they never did. And again, another seed was planted.

Once I wore out my chemistry set, together my parents looked for all opportunities to let me explore these new science interests. This included enrichment camps like the Girls + Math Camp at Western Illinois University, an Engineering Camp for Minorities at Valparaiso University, and a Science and Math program at the Illinois Math & Science Academy. When I think back, camps and activities like those were so critical—because they exposed me to topics in advanced math, areas of engineering, and concepts like problem-based learning. These were things that I was not exposed to in my local school. They also let me see other kids my age who were also interested in these topics—which made it feel more normal to be interested in such things.

Princess went on to study chemical engineering at the Massachusetts Institute of Technology, participating in a myriad of cocurricular activities. She served as the president of the MIT Committee on Multiculturalism, President of the MIT chapter of the American Institute of Chemical Engineers, and held both chapter and zone offices in the National Society of Black Engineers. She pursued her love of the arts, acting in a play ("The Colored Museum"), singing in the concert choir, and taking courses in performance and music: "Not the norm for engineers—but it opened my eyes to many forms of human expression, which was ultimately very fulfilling." Princess also loved athletics, and competed on the women's varsity track and field team, throwing the shot put, weight, hammer, and discus. She set the school record for the 20-pound weight throw and received the 2002 Betsy Schumacker Woman Athlete of the Year Award.

As an undergraduate, Princess engaged in biomedical engineering research and applied the principles she was learning to problems that significantly affect human health. Her research earned her the prestigious Class of 1972 Undergraduate Research Opportunities Program award, presented annually to the project that most improves quality of life through its impact on people and/or the environment.

"Having these research experiences early on—in high school and college— were critical to my pursuit of a career in bioengineering," she says. "As a result, I try my best to provide similar opportunities to undergraduates here at Illinois." She is focused on preparing the future generation of bioengineers at Illinois to be leaders statewide, nationally, and globally, and to encourage them to pursue career options that she views as wide open.

"Illinois Bioengineering graduates can serve patients, engage in biological, medical, and engineering research. They can set science policy, lead research and development in the biotech/pharmaceutical fields, and advance the innovations that can improve our daily lives while fueling the economy."

Princess hopes to diversify that next generation. She and bioengineering colleagues Teaching Associate Professor Jennifer Amos, and Visiting Research Scientist Kelly Cross have a grant to research the experiences of undergraduate women of color in engineering at Illinois, in order to identify ways to increase and sustain their participation in STEM fields of study. Kelly Cross explains, "The grant highlights the unique perspective that women of color bring to the field of bioengineering, but who may have to overcome the double bind (i.e., being the gender and racial minority) common within the STEM culture. The study will enhance our understanding about how gender can be experienced differently by various racial and ethnic groups." Words from their proposal echo Rosalyn Yalow's message in Stockholm: "Women of color represent a tremendous untapped human capital and could further provide a much-needed diversity of perspective essential to sustaining technological advantages and promote positive academic climate."

The outlook is promising, then, for bringing women students and faculty to the nation's first college of medicine based in engineering, coming soon to the University of Illinois at Urbana-Champaign. In addition to the traditional medical school curriculum of basic life sciences and clinical experience, students also will take engineering courses that will expose them to topics such as healthcare systems engineering, imaging and sensing, nanotechnology, cell and tissue engineering, biomechanics and prosthetics, and synthetic bioengineering. Medical school students and faculty will work side by side with their counterparts in engineering to develop new diagnostics, devices, and treatments.[33]

Rosalyn Yalow's words at the Nobel banquet are fitting for this new partnership between medicine and engineering at Illinois: "We bequeath to you, the next generation, our knowledge but also our problems. While we still live, let us join hands, hearts and minds to work together for their solution so that your world will be better than ours and the world of your children even better."[34]

Touching the Sky

ANGELA S. WOLTERS

For years, the department of Mechanical Engineering at the University of Illinois offered an aeronautical engineering option, but eventually, in 1944, Aeronautical Engineering formed its own department. By the late 1960s, it had grown to be the largest Aeronautical Engineering Department of any university in the country, bringing more than six hundred undergraduate students to Urbana. Since that time, research in aerospace engineering has grown to touch more than thirty departmental and interdisciplinary laboratories across campus.[1] The Department of Aerospace Engineering, as it's now called, continues to thrive.

Aeronautical Engineering owes part of its striking growth to the Soviet Union's launch of the artificial satellite Sputnik in 1957. Sputnik marked the beginning of a "Space Race" between the Soviet Union and the United States, a race for supremacy in spaceflight that lasted approximately eighteen years. The race between these Cold War rival countries focused on "gaining the upper hand in science and technological innovation" and was rooted in the concurrent missile-based nuclear arms race—a serious national security concern for the United States.[2] In response to the Soviet Union's efforts, the United States launched its first satellite, Explorer I, in January 1958; that same year, Congress approved the establishment of a civilian space agency to oversee U.S. space exploration efforts. That agency was named the National Aeronautics and Space Administration (NASA).[3]

During the years of the "Space Race," women made up nearly 30 percent of the total engineering workforce in the Soviet Union, versus 2 percent in the United States.[4] The United States made calls for additional engineering manpower—or womanpower, as the case may have been. White men held the vast majority of positions at the time; women were an untapped resource.[5] Despite their small numbers, American female aerospace engineers brought us trailblazers like Barbara Crawford Johnson, who helped the United States ultimately win the "Space Race" by landing a man on the moon on July 16, 1969. Barbara's work on space exploration projects like the Apollo Lunar Landing paved the way for other women—like Victoria Coverstone and Hui Lin "Winnie" Yang—who dream of touching the sky.

✳ Barbara Crawford Johnson (1925–2005)

Barbara Crawford Johnson, who went by "Bobbie" most of her life, wanted to fly from an early age. Bobbie deeply admired Amelia Earhart. As a young girl, she cut her hair in a short Amelia-style haircut and wore a replica of her flight cap. Whenever she could, Bobbie would head out to a dirt landing strip near her hometown of Sandoval, Illinois, to watch the planes and to beg for rides. When she got one, she'd interview the pilot to learn the "hows" of flying and would ask for a chance to guide the plane. As the years passed and she learned the basics, her love for flight grew.[6]

The daughter of a school superintendent, Bobbie described her family as "very education-oriented"; she always assumed she would go to college. But she didn't land on a clear college major until high school, where she realized that her combined love for math, science, and flight could make a perfect entrée into aeronautical engineering. Her ambition was also nurtured by an unexpected role model: a practicing engineer who oversaw the construction of a new high school in her hometown. Bobbie's father was in charge of the school, and their small town had no hotel, so the engineer stayed with Bobbie's family.[7]

Noticing Bobbie's interest in the floor plans for the new building and the slide rule he used for his calculations, the engineer taught Bobbie how to read the plans, introduced her to engineering language, and even taught her some surveying. This experience, along with the exposure to engineering she received from her brother, who studied electrical engineering at Naval Academy, solidified her interest in the field. "I didn't really know at the time that girls didn't take it," Bobbie noted, reflecting on her choice of major. "My dad never said anything: he thought it was a great idea." Her mother did, too.[8] With full support from home, she was on her way.

Bobbie was the youngest of six children. Her father, concerned about the cost of college, urged Bobbie to take a variety of scholarship tests. In the end, the University of Illinois was the most economical choice given the in-state tuition and scholarships. Bobbie enrolled in 1943. At that time, Illinois was a year away from offering an aeronautical engineering degree, so she selected general engineering as an alternative, which provided her with a strong technical background in all engineering disciplines.

Bobbie's transition to college was not easy; she suffered from homesickness and received comments from various people that since she was a woman she would probably have trouble getting a job in engineering. Worried, she considered transferring to business and asked her father for advice. He said, "Don't worry about the job. Just take what you like and if you like engineering, that's what you should stay in."[9] Her mother didn't want her to even consider transferring. Bobbie took their advice and stayed put.

Bobbie still loved to fly. In her free time, she'd frequent the coed student dances hosted at Chanute Field, the Army Air Corps base in Rantoul, Illinois. World War II was on, and the base was active. There she'd "get acquainted with some of the officers who had airplanes at the University of Illinois Airport"[10] and ask them to take her out flying. With their guidance, she learned to fly on her own. But her flight record was not without incident. One summer, when flying one of the officer's planes, Bobbie ran out of gas and had to gently land the plane on a tree in an orchard. It didn't do too much damage, and a local farmer helped to remove the plane from the tree. The incident didn't dissuade her from looking for more opportunities to fly.[11]

While on campus, Bobbie felt well supported by professors, advisers, and fellow students. "Professor Spring was my counselor," she noted, "and he was very, very encouraging all the time . . . Professor Vawter encouraged me tremendously."[12] She also won a bid to become an at-large student senator: "I got more votes than the person running for president . . . [as] the whole college of engineering got out and voted for the first time."[13]

With the support of good educators and a desire to reduce the cost of her college education, Bobbie enrolled in approximately twenty hours of courses each semester to complete her degree in three years. In 1946, she became the first woman to graduate with a general engineering degree from the University of Illinois.[14] Over two other job offers—staying at Illinois to teach and pursue graduate work or heading out east to build bridges—Bobbie decided to pursue her love for aviation through a job with North American Aviation in California (later to become the Space Division of Rockwell International). She feared she wouldn't know enough for the job, but her father assured her that they would provide her with the training she needed.[15]

Despite her general engineering degree, Bobbie's first job assignment was in an aerophysics lab with the aerodynamics group. "I have no idea why they put me where they did," she recalls. "I really didn't even know what a Mach number was."[16] But her first work assignment was very basic—multiplication and division calculations on rows and columns of numbers using her slide rule.[17]

This task did not excite Bobbie, so she met with her supervisor. "I am an engineer," she told him. "I am proficient at the slide rule and I know how to use it very well, but I need no more improvement, and I want a real engineering assignment." At this, her supervisor took his feet off of his desk, started laughing like crazy, and walked out of the room. "Oh, my God," she thought, "I've been fired."[18]

Soon the supervisor returned with the manager of the department. "I understand you want a real engineering job," the manager said. Bobbie said, "That's right. I thought that's what I was hired for. I am a real engineer." The manager shared that he had a challenging assignment for her—working on supersonic inlet design for a ramjet. This shift to working on projects with the conceptual and preliminary design aerodynamics group had a large impact on Bobbie's career. "I got to work with some of the neat preliminary designers who were great at the time. Really, they taught me an awful lot. In a big company, you can get lost in the fog if somebody doesn't know what you want to do. I think you've got to tell them."[19]

Bobbie loved her new assignment and asked for more technical work, which led to other interesting early career opportunities, including "flight dynamic studies for programs such as Dyna-Soar,[20] the recovery of hypersonic gliders, lunar reentry vehicle research, and orbital rendezvous."[21] To increase her background knowledge in these areas, Bobbie took graduate courses in aeronautics at UCLA while continuing to work.

Within five years, Bobbie moved from her original position as a mathematician to a senior aerodynamics engineer and worked on the Navaho missile program, one of the country's first missile efforts. Technologies developed during the Navaho program provided a basis for the Hound Dog missile project, for which Bobbie served as the project lead on "wind tunnel programs, performance and stability analysis, and aerodynamic loads."[22] Her technical contributions were part of the design contracts that Rockwell had with the Air Force.

During her work on the Hound Dog project in the 1950s, Rockwell held a large multiday meeting at Wright-Patterson Air Force Base to update the Air Force on the progress of the project. At this time, project travel was restricted for women. Rather than brief other engineers on the work she was leading so they could report the progress, Bobbie submitted a travel authorization as "B. C.

FIGURE 6.1 Barbara Crawford Johnson, ca. 1963. Photo courtesy of the University of Illinois Archives at Urbana Champaign.

Johnson." It was approved. So without the program manager's full knowledge, she made the trip and gave a briefing so well-received that a high-ranking Air Force official thanked the program manager for sending Bobbie. On her return, Bobbie received a new "travel authorization form for women" that needed authorization by the president or senior vice president of the division. But her bold move broke the barrier for women to participate in on-site client meetings. "I had more travel than I wanted after that!" Bobbie recalls.[23]

During the same period, Bobbie and her husband had a son. For about three years, Bobbie worked three days a week to allow her more time at home. It was a fabulous time for her and her son, but when her team began to prepare their bid proposal to NASA for the Apollo project, she decided to return to work full time. Bobbie was captivated by the thought of sending a "man to the moon." "The Rockwell team worked night and day on this [proposal]," she remembers, "just making everything come together."[24] Once NASA selected Rockwell for part of the Apollo project, Bobbie and the rest of the team started receiving widespread notice for their contributions to manned space flight.

Her work on the Apollo mission focused on performance and design trajectories for atmospheric entry. She also helped lead the development and testing of the entry monitor system, which used graphical display systems installed within the spacecraft for the first time. She and the design team had to convince many naysayers—including some of the astronauts—that it was a good option. "We continued to really sell it," Bobbie explains. "We got it in the [flight] simulator, and we got the astronauts into it and they liked it."[25]

In 1968, Bobbie took on the role of manager of Mission Requirements and Integration for the company's Apollo Lunar Landing contract with NASA, making her the highest-ranking woman in her division at Rockwell. Bobbie's involvement in NASA's manned space flight efforts continued with the Skylab and Apollo-Soyuz (joint USA-USSR) programs,[26] where she enjoyed being "involved in all facets of the design and development" of systems to meet design requirements.[27]

Throughout thirty-six years of work in the aerospace industry—twenty-five of those as a manager—Bobbie led talented technical teams while working on cutting-edge projects. Bobbie's teams often included ninety to one hundred people, and during the Apollo project, personnel in her group swelled to about one hundred sixty. She focused on the expansion of her group's technical capabilities by properly training those around her so that there was always an "up and coming" leader ready to take her place. This management style also allowed her to take on new roles and challenges so her job would never become boring.

A fellow of the Institute for the Advancement of Engineering (IAE) and the Society of Women Engineers and an Associate Fellow of the American Institute of Aeronautics and Astronautics, Bobbie received a number of prestigious awards during her career in the aerospace industry. In 1970, NASA presented her with a plaque for her "outstanding contributions" to the Consumable Analysis Team for the Apollo 7 through Apollo 11 missions. NASA also recognized Bobbie in 1973 with a letter of commendation from astronaut Fred Haise Jr. and a medallion commemorating her contributions to the Apollo 11 mission. In 1974, the Society of Women Engineers honored her with an Annual Achievement Award.[28] The College of Engineering at the University of Illinois named her a 1975 Distinguished Alumni—the first woman to receive this award—for "her originality and technical contributions to space flight dynamics of missiles and manned vehicles, for her unusual capacity to be a contributing member of a team as well as an inspiring leader able to manage both people and projects."[29] The American Astronautical Society granted her the Dick Brouwer Award in 1978 to honor her "significant technical contributions to space flight mechanics and astrodynamics."

For Bobbie, out of all her work, the Apollo Lunar Landing stood out as a contribution "you wouldn't dream you could do . . . [but] we did it!"[30] Her "can-do" attitude and her technical expertise have secured her spot as a pioneer of the U.S. missile and space program, helping humankind touch the moon for the first time.

✳ Victoria Coverstone (1963–)

With Katherine Carroll, Aerospace Engineering Student

While Bobbie was making history with the Apollo space missions, Victoria (Vicki) Coverstone was an aspiring young engineer. During the mid-1960s, as Bobbie and other outstanding engineers were helping NASA's efforts to venture further into space and science fiction shows such as *Star Trek* flooded television screens across America, children like Vicki believed that space travel was the world's next big achievement. An avid admirer of Captain Kirk, Vicki wanted to make a great impact in the field of astronautical engineering.

A late baby boomer, Vicki was born in 1963 to two primary-school educators. Vicki challenged traditional gender roles at an early age by becoming the first girl to play on the Tuscola (Illinois) Little League baseball team, and she continued to resist predictable pathways for women when she entered the University of Illinois as a math and computer science major. She chose this major at the suggestion of her high school guidance counselor. But early in her college career, Vicki questioned it.

She resolved her uncertainty on a coed canoe trip at Turkey Run State Park in Indiana. While canoeing, Vicki chatted about her major with a boy who happened to be an engineering student. He told Vicki about the fields of aeronautical and astronautical engineering, that they were areas you could major in. Their conversation, paired with Vicki's longtime interest in space travel and *Star Trek*, led her to consider switching to an engineering discipline.[31] "I figured that even if I didn't become an astronaut," she explains, "I could [still] be involved in the design of space missions. An engineering degree would set me up with a win-win situation!"[32]

After transferring into the College of Engineering, Vicki faced more challenges, among them the lack of female aerospace engineering faculty and fellow students. "I was the only girl in many of my classes," Vicki recalls. "I sometimes felt that it would be easier to give up, but I took it as a personal challenge to graduate with my engineering degree."[33] Then there was the tricky territory of how to handle her male TAs: "I was asked on a date by my TA in one of my laboratory classes and felt like I had to go so that my grade would not be affected. He turned out to be a nice guy, but the rest of the semester was very awkward."[34] Still, despite the challenges, Vicki knew she had made the right decision to switch majors.

In the spring of 1985, Vicki completed her undergraduate education and was offered a promising job working on space mission design at NASA's Jet Propulsion Laboratory (JPL), but she declined and chose instead to study for her master's degree at Illinois. As a research assistant (RA) under Professor Lee Sentman, Vicki worked in a lab focusing on space-mission design alongside David Carroll, another RA under Sentman and Vicki's future husband. Later, Vicki began working for another adviser, Professor Thomas Dwyer. She completed her master's degree in 1986 in spacecraft attitude dynamics and control.

Dwyer asked Vicki to continue working toward her doctorate. Feeling obligated, Vicki complied, but all the while dreamed of working in California. The job offer from JPL had long since expired, but she'd been offered another tempting position at TRW Space Park in Redondo Beach, California. Longing for a change of pace, Vicki dropped out of her PhD program in October and accepted the offer from TRW, who also hired David. Together they moved to California and married soon afterward.

After about eighteen months of living in California, Vicki realized that her ultimate career goal was to become an astronaut. She always had an interest in space travel, but until her time at TRW, she did not seriously consider the possibility of turning that dream into a reality. Researching the job requirements, Vicki found that a candidate must have at least an MS degree plus three years of relevant work experience or have a PhD. After discussing their options, Vicki and David concluded that TRW could no longer support their professional goals. In 1988, they returned to Illinois, where Vicki continued working with Professor Dwyer for about a year until his untimely death in a car accident.

Following Dwyer's passing, the department asked Vicki if she would cover the course that Dwyer had been scheduled to teach. Feeling that it was the least she could do, Vicki promptly accepted the offer and began preparing for the class, "Aerospace Control." "It was my first experience teaching," she remembers, "and I loved it!"[35] It was the first of many classes that she would teach over the course of her career.

Vicki continued her doctoral studies with John Prussing, professor emeritus in the aerospace department and obtained her PhD in aeronautical and astronautical engineering in January 1992. That spring, the Aeronautical and

Astronautical Engineering (AAE) Department announced that they were looking to hire a new faculty member. Vicki decided to apply, and she was offered the position of assistant professor; David was also offered a full-time position as a research scientist. Vicki became the first woman faculty member in AAE at Illinois.[36]

Early in her career as a professor, Vicki applied for a position as astronaut. Many people dream of becoming an astronaut, but few make it to the interview stage. To her astonishment, Vicki was invited to Houston, Texas, for a personal interview in August of 1994. There she learned that, of the approximately 2,000 applicants for the position, NASA was interviewing 122 qualified individuals and would hire approximately 25. Vicki wrote in her diary at the time, "The odds are still steep, but I always like a challenge."[37]

The weeklong selection process included long and tedious medical examinations that tested Vicki's physical and mental health. After interviewing at NASA Johnson Space Center and meeting many inspirational aerospace professionals, such as Dan Goldin, the NASA Administrator chosen by President Bill Clinton, and Story Musgrave, NASA astronaut, Vicki was medically disqualified because of issues with her vision. "The hardest part about not making it," she recalls, "is feeling inadequate."[38]

Although difficult to overcome, this professional disappointment brought clarity to Vicki's career plans. She remembered a time when, as a college student, she had spoken with Steve Nagel, an active astronaut. "Vicki," he said, "you should choose a profession that you are really interested in practicing. Then you will be happy whether you are an astronaut or not."[39] With the additional guidance of her educator-parents, Vicki decided to continue teaching. Her favorite way to spend time was with "young, bright people who are interested in learning about space."[40]

In 1998, Vicki became the first woman to receive tenure in AAE when she was promoted to associate professor. She went on to become full professor in 2005. She won several teaching awards, including the Everitt Award for Teaching Excellence, three-time Teacher of the Year in the Aerospace Department, the College of Engineering Teaching Excellence Award, and the Campus Award for Excellence in Undergraduate Teaching—Honorable Mention. Later in her career, Vicki won the College of Engineering Stanley H. Pierce Award for outstanding undergraduate teaching. And she has been named as an outstanding faculty adviser for both undergraduate and graduate students.[41]

It shouldn't be surprising that such a gifted teacher would be an equally gifted leader. Vicki's leadership has enabled many of her undergraduate students to win multiple American Institute of Aeronautics and Astronautics (AIAA)

National Space Competitions. A fellow of the AIAA and a member of other professional societies, Vicki has also chaired the Council of Institutions for the University Space Research Association.[42]

Vicki is also an influential businessperson. In 1998, a number of the aerospace professors at the University had ideas for potential commercial products based on past research. Vicki and David, along with four other professors, started a small business called CU Aerospace. As cofounders of the company, David was the first full-time employee and Vicki secured the first contract: with Orbital Science Corp to perform optimal maneuvers with OrbSat satellites.[43] Since its founding, Vicki has remained active with the company, writing many of its spacecraft and astronautics proposals. Over time, CU Aerospace has added new partners, and two of the company's largest contracts have been with NASA on solar sailing studies. Solar sailing uses large mirrors to capture sunlight, which then propels spacecraft via radiation pressure.

Throughout her time as a faculty member, Vicki also became interested in working as an administrator, a goal she first met in 2006, when she began serving as the associate head of graduate studies in the Aerospace Department. Two years later, she was appointed associate dean for graduate and professional studies in the College of Engineering, where she worked to develop programs for graduate students. But she missed teaching, so in 2014, she resigned and returned to her position as full-time professor. Watching hardworking students learn and succeed in the classroom remains her favorite part of her job.

One of Vicki's more recent professional accomplishments was being named a Visiting Investigator Partner at the NASA Marshall Space Flight Center near Huntsville, Alabama, in early 2013.[44] She also spent the summer of 2015 there, working on two projects through the Chief Technologist's office. First, Vicki and her team investigated the use of a reflective mirror to deflect some incoming sunlight as a means of counteracting the carbon dioxide (CO_2) buildup in the atmosphere with the end-goal of reducing global warming. Second, they sought to determine whether the use of a ground-based laser to illuminate a solar sail in Earth's orbit would create active propulsion that could be used to accelerate spacecraft. This project compared the effects of concentrating laser light versus conventional solar energy on the solar sail to determine the resulting increase in a spacecraft's propulsion.[45] That fall, Vicki returned to the University with yet more knowledge and experience to pass on to her students.

That same fall brought another rewarding facet to Vicki's professional life: her daughter, Katie Carroll, entered the freshman class in aerospace engineering at Illinois. Vicki is excited about having Katie on campus: "Katie possesses the intelligence and passion required to carry forth the progress that women

FIGURE 6.2 Victoria Coverstone. Photo courtesy of UI News Bureau.

have made in STEM fields. I'm looking forward to collaborating with her as her professional knowledge grows."[46] For her part, Katie admires her mother's hard work and determination: "I am eager to begin my studies in aerospace, and am enthralled by the idea that I may study this discipline under my own mother, who has always been a source of motivation."

Like Vicki's students, Katie, too, learned valuable lessons from her mother: "Of the many things my mother has taught me, one is to know that it is great to be determined to go after certain goals, but it is also necessary to understand that other possibilities can arise if those goals do not turn into a reality." Vicki's flexibility, perseverance, and aspirations to reach for her dreams are just some of the pivotal reasons why she serves as a role model for young women pursuing math and science. For Katie, her mother's story has shown her "that the sky is *not* the limit."

Katie was also inspired to pursue aerospace engineering by attending GAMES (Girls' Adventures in Math, Engineering, and Science) Camp in 2014 on campus at the University of Illinois. "I enjoyed this experience," she said, "because throughout my life, I had always seen the 'astro' side of aerospace from my parents. This was the first time that I also got to study the flight of airplanes, and I ended up loving both!" Katie knows now that no matter what field she ends up concentrating in, she'll be doing work she enjoys, and that, she says, "is all that really matters."

Since the beginning of her own education at the University of Illinois in 1981, Vicki remained an integral part of the Aerospace Engineering Department throughout most of her professional career. During that time, she has seen a growth in opportunities for women in engineering. Yet the aerospace industry in the United States still lags in its numbers of women aerospace engineers. Women represented nearly 20 percent of all undergraduate engineering degrees throughout the United States in 2014, but only 13.8 percent of aerospace engineering degrees went to women.[47] Luckily for the aerospace industry, the interdisciplinary nature of the work has brought many women to aerospace from different engineering backgrounds, including but not limited to mechanical engineering, electrical engineering, civil engineering, and materials science and engineering.

Hui Lin "Winnie" Yang, a 2015 graduate in materials science and engineering at Illinois, found alignment between her technical interests and the aerospace industry. Here, she shares her reflections on her career as an engineer with Boeing, as well as on the future of women in aerospace engineering.

✳ Reflections by Hui Lin Yang

BS Materials Science and Engineering, 2015
Materials and Process Engineer, Boeing

I wanted to work in the aerospace industry since the summer before my senior year of high school. I was attending an engineering camp, and one of our visits that summer was to the Boeing facility in St. Louis. We toured the production of fighter jets, and I fell completely in love with the idea of creating these giant machines that have the ability to move people across and even out of this world. This was definitely the field I wanted to have a career in. I didn't know exactly how, but I had found my calling among the planes.

The next fall, I started my undergraduate career at the University of Illinois in Materials Science and Engineering (MatSE). I moved in almost a week before classes actually started to be a part of the Women in Engineering (WIE) Freshmen Orientation camp, which in later years was sponsored by Boeing. Not only did coming a week early allow me to move into my

Details of the Women in Engineering (WIE) Freshman Orientation are described in detail in the section, "The Following Fifty Years: 1967 to 2017," in chapter 1, "Engineers Who Happen to Be Women." In recent years, companies such as Abbott, Boeing, and Texas Instruments provided the resources necessary for the successful execution of orientation.

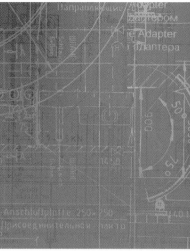

FIGURE 6.3 Hui Lin Yang. Photo courtesy of Carla Estridge.

residence hall before the big rush of students, it also provided the opportunity to meet other women engineers and get acclimated to the campus before all the students came. Without the camp, I would not have made as many friends as I did—friends who would later become my study buddies, my roommates, my mentors, and my partners-in-crime.

Armed with these resources, I sought out other opportunities in areas that interested me. Throughout college, I was an active member of the Society of Women Engineers (SWE) and took part in helping organize professional events with other societies and events that reached out to K-12 students in the surrounding area. The professional events provided a way to interact with companies and build relationships with corporate sponsors. The outreach events provided a channel for younger students to explore STEM and learn about its different possibilities. In addition to SWE, I was a part of the Hoeft Technology and Management Program (T&M), a privately sponsored minor that allows business and engineering students to integrate our experiences and learn from each other.

The Boeing Company had a steady presence in both SWE and the T&M program, acting as a corporate sponsor in both organizations. As a University of Illinois alumna, Dr. Dianne Chong—vice president of Materials, Manufacturing, Structures and Support—championed many of the events and became a sort of guiding light for me. So many of her accomplishments were ones that I hoped to make in my career; I aspired to be like her. She had found a way as a metallurgist to work at a plane factory, and I was determined to do the same with my materials science degree. I did not want to be closed off from an area of industry just because the title of my engineering degree didn't seem to fit. It was important to me to work in a company that would use my talents to advance their current capabilities rather than stick to the status quo. Sure, I would have to learn the technical jargon of aerospace to keep up with others, but I think the best learning is done on the job and not from a book or in a classroom. It was a challenge I accepted with open arms.

All that leads to the present. After two internships with Boeing, I started full-time with them in August 2015 as a Materials and Process Engineer within the Next Generation Composites group. The Boeing Company celebrated one hundred years as a company in 2016, and it's breathtaking to look back and see what has happened in that time, from the production of wooden airplanes to sending people to the moon. I currently work on a project that is aiming to send astronauts to the International Space Station and back using some new materials that have never been used in space flight before. The next hundred years have the potential to take ordinary people to farther planets and bring them back, as well as to change the landscape of leisure travel beyond anything the Wright brothers had ever imagined. Through advances in the types of material systems and innovative processing capabilities, I can confidently say that I am creating the aircraft of the future.

Women's Work

Computing

ANGELA S. WOLTERS

The indispensable machine we know as the computer got its start at the University of Illinois in 1952. That early automatic, electronic digital computer—the first of its kind owned and operated by a university—was known as ILLIAC I, which stood for Illinois Automatic Computer. ILLIAC I was built at Illinois along with a machine named ORDVAC (Ordinance Discrete Variable Automatic Computer), which was designed and constructed for the Army's Ballistic Research Laboratory. ILLIAC I began operations on September 22, 1952, at the start of the fall semester.[1]

The presence of ILLIAC I engaged students in programming and computational work and began a long, storied history of computing at the University of Illinois, which includes the contributions of many women.[2] From those early pioneers like Ramona P. Borders (Figure 7.1), who worked with the first machines on campus, to those working today in the high-tech sector, these women and their stories inspire students interested in pursuing computing careers.

The engineers who designed the first computers—almost entirely men—believed that developing the hardware was the challenging task and that programming required primarily clerical skills, which fit squarely in the arena of "women's work" at the time. So they hired women programmers—mainly recent college graduates with math degrees.[3]

FIGURE 7.1 Ramona P. Borders with ILLIAC. Photo courtesy of the University of Illinois Archives at Urbana Champaign.

What the design engineers didn't realize was that programming these machines was as challenging as building them. In fact, the ILLIAC I Programming Guide, published over a year and a half after ILLIAC I became operational, notes these challenges specifically:

> The occurrence of cycles is one of the things that complicates the programming of a calculation. Another is the fact that, since orders are stored in the memory in the same form as numbers, they can be operated on and altered during the course of a calculation (at the behest of other orders) just as if they were numbers. All this makes possible some most interesting calculations; it can also make programming difficult.[4]

While some of the programming work was more clerical, much of it required deep knowledge of logic, mathematics, and the electrical circuits that made up the "brains" of the machine. As women coded these complex routines, others looked on and recognized that the work associated with computer programming was in fact "a high-level, challenging, and creative occupation." Over time, as full knowledge of the level of logic needed for this programming work unfolded, men decided to pursue these jobs, replacing women and dominating the computer programming workforce.[5]

✳ Nancy Brazell Brooks (1930–2008)

One of the women involved in the early efforts of programming ILLIAC I for research was Nancy Brazell Brooks (MS 1953), a graduate student researcher in the Department of Civil Engineering. Early on, Civil Engineering was one of the two largest users of ILLIAC I on campus.[6] The department describes Nancy as "one of the pioneers in bringing computers into use in the department."[7]

Nancy earned a bachelor's degree in architecture with concentrations in structures and design from Alabama Polytechnic Institute (to become Auburn University in 1960). The dean of the school of architecture had tried to convince Nancy to forgo the study of structures, as it was "unladylike." But Nancy ignored him. She knew better. Even as a child, she had neglected "her dolls in favor of her brother's Tinker Toys and Erector Set."[8]

After graduation, Nancy worked a short while for an architecture firm. But the Korean War caused a shortage of structural steel, which slowed construction across the country, leaving little business for architects. So Nancy left the firm and decided to attend graduate school. Her former professor of structures told her Illinois was "the only place to go." Nancy secured a research assistantship to work with Nathan Newmark and became the first woman on the department's graduate student research staff. This first was not without challenges. "They didn't know what to do with me," she admitted. "They didn't know what office to put me in. In fact, they had a meeting for the sole purpose of determining that."[9]

Nancy also noted that she was not allowed in the experimental crane bay—the central Civil Engineering laboratory for large-scale experimental research—but "was allowed to expand [her] talents in the use of computers for structural problems." The crane-bay ban would ultimately prove beneficial for Nancy's future, as she instead developed computer-programming skills using ILLIAC I that "only a small number of people nationwide" had.[10]

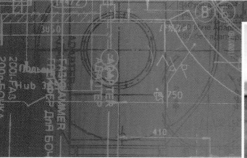

FIGURE 7.2 Nancy Brazell Brooks. Photo courtesy of the Civil and Environmental Engineering Department, University of Illinois.

Upon graduation in 1953, Nancy married her Civil Engineering officemate, John A. Brooks. Due to university nepotism rules that prevented spouses from working in the same department, Civil Engineering had to request special permission from the provost's office to offer Nancy a position as a research associate.[11]

From 1953 to 1955, Nancy led research work for the Physical Vulnerability Division of the Directorate of Intelligence of the United States Air Force. She developed computerized methods to determine the dynamic response of structures to air blasts and ground shock and summarized the work in two reports released to the Air Force in 1954 and 1955, respectively.[12] While her work was conducted under the general direction of Nathan Newmark, the future Civil Engineering department head, Nancy was the primary author on both reports.

In the 1954 report, Nancy acknowledges her research assistants for collecting the reported data. One of these assistants was Dolly L. Marsh (Gurrolah). Dolly went on to serve as a secondary author on the report published in 1955. Nancy and Dolly had received their bachelor's degrees in architecture two years apart at Alabama Polytechnic Institute. We are left to wonder how much Nancy and Dolly interacted prior to or after their work together at the University of Illinois. At minimum, Nancy supervised Dolly as she completed her master's research, and Dolly became the second woman licensed as a structural engineer in the state of California in 1962.[13]

Nancy's work with computers at Illinois influenced the rest of her career, including her first position working as a mathematician in the computer science department for RAND Corporation. Located in Santa Monica, California, RAND had a sister computer to ILLIAC I, which not only made Nancy a perfect fit as an employee but also allowed her to get involved in some very interesting projects early in her career. In an interview conducted in 2003, Nancy shared details of one of her first work assignments: "One of the first ones—now, this was 1955, long before Kennedy said we were going to the moon—was to determine what entry velocity and in what direction to aim to impact the moon from earth. I hit the moon on the third try!"[14]

She left RAND after eight years to cofound a research and development firm, Shock Hydrodynamics Corporation, Inc., which was later acquired by Whittaker Corporation, a Fortune 500 company in the '60s, '70s, and '80s. Later, as a senior systems consultant with General Research Corporation, Nancy worked on government projects involving structures, the dynamics of structures, and computer application of engineering models. She retired after more than twenty-five years of engineering computation contributions. During her retirement, she was very active in charitable work.

> Another woman in the College of Engineering who utilized ILLIAC for her research was Geneva Belford. Her work focused on programming to address the chemistry topics of radiation kinetics and sedimentation of reacting systems. Read more about her work and how she served as a mentor to one of the biggest names in technology in chapter 10, "Mentors and Mentoring."

In 1998, Nancy won the first Distinguished Alumna Award from the Civil Engineering Department. Before her death in 2008, she and her husband established a graduate student fellowship in Civil Engineering.[15] With this gift, Nancy continues to advance engineering practices just as she did throughout her notable career.

✳ Hillery Hunter (1978–)

(Adapted from an essay by Jonathan Damery[16])

While Nancy and Geneva represent some of the first women in computing at the University of Illinois, many others have played seminal roles in advancing computing technologies through their professional practice. One of these is alumna Hillery Hunter (BS Electrical Engineering (EE) '99, MSEE '02, PhD '04).

Hillery thinks about engineering the same way that she, as an accomplished pianist, approaches the ebony and ivory. It's a counterpoint of roles, always planned several phrases ahead and performed across the full tonal range—spanning, in engineering, from software to market analysis, from the conceptual to the applied.

As director of Accelerated Cognitive Infrastructure research and an IBM Fellow, Hillery leverages technical expertise in system design, computer architecture, and memory to help define strategy for future IBM systems. Hillery leads a team doing architecture, algorithms, and system projects for cognitive computing—systems that excel at machine learning and deep learning. She's responsible for making sure that IBM research projects on a broad set of topics—from next-generation computer memory to accelerators—computer elements that are optimized to perform particular tasks efficiently—move and develop in concert. "It's a very broad role," she said, "keeping the research teams across everything from silicon to software interlocked with our development organization . . . so that we're sure that we're working on things that are going to push forward the state-of-the-art for IBM systems."

IBM is an industry leader for high-bandwidth, high-capacity memory systems, and Hillery masterminds a portfolio of projects related to these, envisioning how the systems can better handle computing challenges in the coming years, including the growing demands of big data and analytics. She identified the problem: "How does a business not just have data, but get insights from those data and monetize that information? Memory, including new memory technologies such as flash, . . . is going to be a really key piece of that overall equation for companies." Flash memory is widely used for data storage and data transfer.

Hillery started at IBM in 2000 as an intern at the IBM Development Lab in southwestern Germany, a position she earned based partly on her experience studying engineering abroad as an undergraduate at the Technical University of Munich. Having become essentially fluent in German during two previous summers in the country, Hillery proved her aptitude for both German and engineering by winning the University of Illinois' Outstanding International Student Prize. "Having [those two semesters abroad] greatly enhanced my experience at Illinois," Hillery said. "I work now with a team that is spread everywhere from California to Zurich to Tokyo to Korea. We work globally. We do proof of concept projects with customers around the globe. . . . So that experience of being somewhere else, dealing with a different culture, while still being in a technical environment, has been invaluable."

After three additional internships at the company's T. J. Watson Research Center, Hillery joined IBM full time in 2005 as a member of the exploratory computer systems group. There she argued for the cost-performance benefits of incorporating embedded dynamic random-access memory (eDRAM) onto

FIGURE 7.3 Hillery Hunter, IBM Fellow. Photo courtesy of Jon Simon/ Feature Photo Service and IBM.

IBM's processor chips. Although eDRAM has a higher cost per bit, its performance outshines that of the lower-cost true DRAM equivalents. Based on her work, which drew from research in various departments, IBM equipped all of its 2010 processor chips with eDRAM.

IBM then asked Hillery to lead a project relating to memory power and the use of DDR3 (double data rate type 3) memory. Because IBM does not manufacture this memory, she managed interactions with external vendors, establishing specifications for the products, as well as the initial, in-house design initiatives that aimed to lower the power consumption of the devices. "My work," she explained, "was largely in getting things translated across technical teams and really making sure that technically we had good communication flow and that we were making the right decisions based on what other teams were doing."

As an undergraduate and graduate student in electrical engineering, Hillery worked in the Coordinated Science Laboratory with her graduate adviser,

Professor Wen-mei Hwu, a leading authority on computer architecture, as well as with Professor Naresh Shanbhag, whose expertise lies in integrated circuits design and signal processing. "The interdisciplinary nature of centers such as the Coordinated Science Laboratory is really important in preparing students for the real world," Hillery said. "In computing in general, we are running into problems that require people with interdisciplinary understanding and [a] willingness to tackle problems from a different angle."

Hillery earned an IBM Outstanding Technical Achievement Award in 2011 and was admitted into the IBM Academy of Technology in 2012. She was appointed as an IBM Fellow, IBM's highest technical position, in 2017. These honors, along with some 20 other awards from IBM and elsewhere throughout her career, are a testament to her accomplishments in those regions of interdisciplinary overlap, orchestrating disparate parts to act in unison, creating, like fingers on the keyboard, a collective and innovative whole. "Everything I've done here has been interdisciplinary," Hillery said. "Absolutely."[17]

✳ Parisa Tabriz (1983–)

Parisa Tabriz, an alumna of the Department of Computer Science (BS 2004, MS 2006), is the director of Engineering in Chrome at Google. But Parisa is more creative than that title would suggest. When she needed to print business cards for a security conference she was attending in Japan, she chose *Security Princess* as a more fitting moniker. Not because Parisa is "the princess type"—she's practical and down-to-earth—but because it was "cute and a bit more whimsical."[18] There's irony there, too. Parisa proudly acknowledges being a woman in a field dominated by men.

Parisa embraces her passions, from rock climbing to photography, art to gelato making, building elaborate gingerbread houses to leading a team of computer scientists to hack into Google's systems with the goal of providing better online security for the world—undertakings that are far from the typical activities that many would associate with the title "princess."[19]

Coming from two parents with careers in medicine, Parisa knew little about electrical and computer engineering (ECE) when she selected it as her initial college major. But she was skilled in math and science and had been guided toward engineering by her high school teachers and counselors. She reasoned that if she applied to what she perceived to be the most difficult major in engineering and was admitted, she could learn more about the others and eventually find her place. And that's exactly what she did. During her first ECE course, she spent a lot of time in the lab. "I didn't really love it," she noted, "because on

FIGURE 7.4 Parisa Tabriz. Photo courtesy of Brandon Downey.

one of our projects, the wires came loose from the device. I just found it really frustrating to deal with equipment and hardware because things would break on me."[20]

Meanwhile, Parisa's out-of-class interests were shaping her future. She applied for a position as a network administrator in her dorm's computer lab, where she learned about networking and helped others troubleshoot computer problems. She also began to explore web design because "it was an easy way to make art online . . . that made me experiment with programming and made me realize that I wanted to try computer science."[21] After enrolling in her first programming class, Parisa changed her major.

Parisa's role as Google's Security Princess has earned her some impressive attention. *Forbes* included her on its 2013 "30 under 30" list of tech pioneers, *Wired* named her to the 2017 List of 20 Tech Visionaries who are Creating the Future, "CBS This Morning" has profiled her, and she's been featured in the popular media. Even so, she wasn't always confident in her role as a woman engineer. "I wasn't secure in college," she reflects. "I switched majors. . . . Every semester was like, 'Is this worth it? Am I doing this?' And I think women struggle with that [uncertainty] much more."[22]

By candidly sharing her college and career experiences, Parisa hopes to dispel some of the myths she feels exist about studying computer science and finding careers in high tech. "I didn't have this passion or this drive to know what I wanted to do, and I also didn't start programming until my freshman year. It's a myth that you need to have been doing this since childhood." Instead, Parisa urges students to explore the multitude of opportunities on campus and find their passion.

That's what she did: she conducted research, worked as an intern, and got involved with student societies, all of which allowed her to connect closely with computer science. Through the Association of Computing Machinery (ACM), Parisa gained not only leadership skills, but also a supportive community of peers and great senior and graduate student mentors. Even less-than-perfect experiences, such as a summer internship at a pharmaceutical company, helped her gain clarity about her path of working for a company that had software as its central focus.

Parisa admits that her efforts to "find the perfect experiences," while ultimately fruitful, were also sometimes a source of stress, particularly when she switched majors and determined her specialty for her master's degree: "If I could go back and tell myself to 'Just chill,' I would. One of the things I think I've learned postgraduation is that it's a marathon, not a sprint. It is not going to matter what semester you graduate. Take classes that sound interesting, regardless of whether they fit into some thematic area of expertise you feel pressured to build up. College is the time to try new things and optimize for learning."

After graduation, Parisa joined Google as an engineer in the security group. She recounts that after about five years, the group had "grown to a size where we had 16 awesome individual contributors but weren't working as a team. So I wanted to step up to that." She was selected to serve as manager of the group, which included many men who were older and more senior at Google. As an older sister to two younger brothers, she was not intimidated by leading a male-dominated team of hired hackers. The position also allowed her to have different experiences every day, which she knew she needed in a career.

The mission of her team, to make Google Chrome the safest way to browse the internet, is one of Parisa's passions. "As more of people's lives and more of society has become dependent on the Internet, crime and threats have followed. Cybersecurity is not just about protecting someone's credit card number either. In many parts of the world, citizens' online actions and communications are being closely monitored, putting their lives and personal safety at risk. I feel a responsibility to help." Parisa has found fulfilling work at Google leading her team to creatively solve security challenges that will make a difference in the world.

Parisa urges students and those early in their career to "figure out what you like and what you want to get better at and spend your time on that."[23] By all accounts, she has practiced exactly what she preaches.

Today, more men than women are still choosing to study computer science and computer engineering. But the numbers of women are increasing. The percentage of freshmen women enrolled in computer science degree programs at

Illinois tripled from 2009 to 2016, in part because of expanded outreach efforts by the department and the College of Engineering. While the department was delighted that 25 percent of its freshmen students were women in fall 2014, it still aspired for equal representation of women earning computer science degrees at Illinois. In the fall of 2016, the department celebrated a freshman class of 47 percent women while the national average stood just below 15 percent.[24]

Beyond recruiting and retaining women students in computer science, inspiring them is also important. The instructors who influence students daily play critical roles by providing relevant course content, engaging students in meaningful learning experiences, and conveying their own passion for the field. Cinda Heeren, teaching professor in computer science at the University of Illinois (2004–2017), and 2015 American Society for Engineering Education (ASEE) Illinois-Indiana Section Teacher of the Year, is one such individual. Here she shares her reflections on computing and how she will shape her teaching to further break down barriers, not only for women but for all people.

✴ Reflections by Cinda Heeren

PhD Computer Science, 2004
Senior Instructor, University of British Columbia

I've loved computing for decades, but my perceptions of the craft of designing algorithms and writing code have changed in a very subtle way in just the last few years. I had three distinct experiences that showed me, unequivocally, that the roots of my discipline are not owned exclusively by those who popularized it as a commercial and military discipline in the era of World War II.

A few years ago, partly because I missed my Grammy, I picked up my knitting needles and searched for a project of appropriate scope and beauty. What I found instead was a connection to computing that I didn't expect: the language paradigms used to describe knitting patterns and procedures is composed of exactly the same elements as that of C++ or Python or any other modern computing language. Iteration, abstraction, conditionals, and the stitches themselves form a configuration analogous to some kind of data structure. I am not the first to have made this observation, but it has created in me a new sensitivity to the ways in which handcraft is communicated, and most poignantly, it reinforces the idea that computing *constructs* are more general than their application to computing *devices*.[25]

Somewhat later, in the summer of 2013, I spent one day in a textile museum in Guatemala City, Guatemala. My earlier observations about the knitting patterns were reinforced, and I was increasingly touched, because the art of

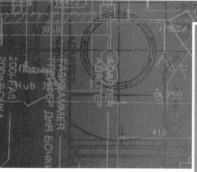

FIGURE 7.5 Cinda Heeren, 2016. Photo courtesy of Laura Hahn.

weaving is hundreds of years old, and yet the handcraft itself illustrates these same computing constructs.[26] Again, the actual act of creating the woven patterns is infused with algorithms and a mathematical understanding of pattern generation.[27]

Most recently, I happened to watch a TED talk from Ron Eglash in which he described the fractal or recursive structure underlying African villages.[28] Recursion[29] is a powerful mechanism for expressing the structure of computing problems. It is applied to optimization problems,[30] and to data structure definitions,[31] and its mastery is considered foundational for any computer scientist. The accepted history of recursion is a journey through Western mathematics and computer science, and yet there are several African cultures for whom the notion of self-similar structures, both social and architectural, has existed for hundreds of years.[32]

The computing of today has its roots in traditions of handcraft, art, and music, and the minute we all admit that fact, we have the ability to break down barriers. With this in mind, I will infuse my teaching with these connections and broaden the historical view of our field, primarily because I do not want artificial barriers to participation. Besides, arts and crafts are beautiful and expressive and *human*.

Global Challenges

LAURA D. HAHN

Energy and the environment, global health, and water resources are among the "critical global challenges" identified by the National Academies of Science, Engineering, and Medicine as priorities for research.[1] And faculty, students, and alumni from the College of Engineering are working on solutions. Tami Bond, professor of Civil and Environmental Engineering, addresses the environmental challenge of air quality. In this chapter, we highlight her work as well as the experiences of women students in Illinois's chapter of Engineers Without Borders, a volunteer organization dedicated to improving the world and empowering communities through engineering.

✳ Tami Bond (1964–)

Whenever something burns, from the diesel in an engine to the kerosene in a lamp, black carbon is released into the air. According to the U.S. Environmental Protection Agency, black carbon emissions are "linked to adverse impacts on ecosystems, to visibility impairment, to reduced agricultural production in some parts of the world, and to materials soiling and damage."[2] They also affect public health, occasionally causing respiratory and cardiovascular disorders. Tami Bond wants to unravel the reasons behind these complex effects and to make a comprehensive case for implementing alternative, less hazardous technologies.

In 2014, the MacArthur Foundation named Tami a MacArthur Fellow. Also known as the "genius grant," this prestigious award honors individuals for creativity, depth, and potential in their work. The intent of these unrestricted awards is to give recipients the freedom to pursue their own agendas. And the spirit of the award suits Tami perfectly.

"I never learned the word 'impossible,'" says Tami of her upbringing. Her parents, whom she describes as "excellence jockeys who never did anything halfway," fueled her passion for learning and her proclivity to explore her interests without constraints. Her mother taught her to read at an early age, and when she was twelve years old, her father let her take his college chemistry class. "He never told me it was weird," she said.

Tami's academic pursuits have followed those themes: Do what's fun and disregard boundaries. For Tami, what's fun is learning how things work. Just after finishing high school, her car broke—"and that was more fun than anything I had done before." After getting a job in an auto shop to learn about fixing cars, she realized there was more to learn about how things work, and she decided to pursue a degree in engineering.

Tami went to the University of Washington for a BS in mechanical engineering, where she learned how to combine "what's fun" with academic rigor. One of her most influential professors, Larry Palmiter, instilled this mindset in her: "You don't understand anything until you do it from the ground up, again and again." Upon graduation, she wasn't interested in getting a job, so she decided to pursue a master's degree at the University of California at Berkeley. She said, "I wanted to do either combustion or atmospheric science to combine chemistry and fluids. I picked my MS program [in Mechanical Engineering] because I could burn things and because I would have a cool advisor."

Tami's PhD work on combustion was, unsurprisingly, a deliberately interdisciplinary endeavor spanning the fields of mechanical and civil engineering as well as atmospheric science. She explains, "It was a perfect way to merge study in chemistry, heat transfer, and thermodynamics, all of which I really enjoyed. I found myself puzzled, though, by some of the disciplinary boundaries. I met people whose funding was based on reducing emissions from combustion, yet their eyes would glaze over when they heard a talk about air quality. While these researchers produced excellent results by focusing on their specialty, I was not satisfied. I wanted to know about the entire lifetime of combustion products, which begin in the heart of the flame, proceed through the tailpipe (or exhaust stack), waft into the atmosphere, interact with other chemicals or with solar radiation, and eventually get changed into another chemical or get stuck on raindrops, the earth, trees, lungs." That sounded like fun to her.

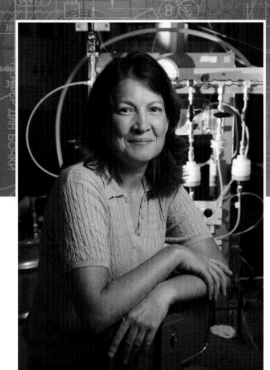

FIGURE 8.1 Tami Bond, 2012.
Photo courtesy of UI News Bureau.

Tami joined the faculty of the University of Illinois's department of Civil and Environmental Engineering in 2003, finding it a place supportive of inter-disciplinary efforts. Her early research involved "equal parts laboratory mea-surements and modeling of particle behavior in the atmosphere, with a dash of humanitarian impact and policy relevance." This interdisciplinary approach—the combination of her technical work measuring black carbon emissions and her focus on the implications of these emissions on society—is what captured the attention of the MacArthur Foundation.

Benito Mariñas, head of the Civil and Environmental Engineering depart-ment, praised Tami at the reception honoring her for her genius grant, noting her contribution toward "understanding the key role played by black carbon in climate and health at the global level." He also applauded her commitment to "protecting the health of those most vulnerable in the world," citing the research she and her students have done on "the development of safe and efficient cook-ing stoves and supporting NGOs that promote their use." Millions of people die annually of smoke exposure from cooking stoves; poor women and children in developing economies are especially hard hit. "But there is hope," Mariñas added, "when they have such a brilliant and dedicated ally."

Tami has taken the MacArthur fellowship as a license to—again—push the boundaries of her scholarship and integrate her research into a broader effort to solve real global challenges:

> I made my reputation on black carbon, but I'm really getting away from it. Measuring black carbon in smoke is interesting but not enough. We also need to figure out how to change it, so I've now done some work on how energy use and air emissions could evolve in the future. Black carbon's blackness, or light absorption, changes the atmosphere, but particles emitted from burning also change cloudiness, so I look at that, too.
>
> Scientifically, I'd like to see the growth and formalization of what I call "anthrogeoscience." I define this as: "The search for causal, mechanistic, generalizable relationships describing the dance between humans and the earth system." I work on the physical side of the dance, but I've increasingly moved toward the human side. Scientists have been working toward understanding physical causality in earth science for a long time, starting with "What happens when the apple falls from the tree?" Now humans are so numerous that they can't be divorced from that system.
>
> I'd also like to see scientific knowledge serving a wider swath of society. A lot of people focus on pushing the leading edge of high tech, which does generate a lot of advances. I'd like us to reach more people who are on the trailing edge. I'm not talking about sending our old computers abroad, which has its own problems, but about making a difference with knowledge that is "good enough." When people are resource-limited, they need robust, tough technology that won't break and will meet a lot of different needs. That's very different from what they are getting now.

Tami and her research group are now focused on gaining a deeper firsthand understanding of the human side of the effects of black carbon produced by cookstoves, brick kilns, and diesel engines. By traveling to areas of the world where these problems are prevalent (primarily in south Asia), she and her team are able to better understand the educational, cultural, and technical challenges of implementing cleaner solutions.

✳ Women in Engineers Without Borders

Ann-Perry Witmer, College of Engineering, Student Chapter Advisor
of Engineers Without Borders

Just as Tami Bond's passion for knowledge led her to tackle global challenges in an innovative, hands-on way, University of Illinois women have been drawn to the opportunity to serve the international community creatively and directly by becoming active in Engineers Without Borders, a student-run organization that flourished on campus after its founding in 2004. The organization quickly attracted a diverse group of volunteers from throughout the engineering campus, and many of these volunteers point to Engineers Without Borders as pivotal in helping them to define their own identities. Women in particular have found a voice as they filled significant leadership roles in the organization since its inception, but no one in the short history of Engineers Without Borders emerges as a stronger, more revered, or more influential leader of other students—especially of other women—than Maren Somers.

Maren was an energetic sophomore in Civil and Environmental Engineering when T. Patrick Walsh (BS Engineering Physics, 2007) first met her at an information night for the newfound Engineers Without Borders chapter on campus. The poster he saw in Loomis Lab drew him to the meeting, but what kept him involved in the organization was Maren.

"She was just a unifier of people," recalls Patrick, founder and chief technical officer of Greenlight Planet Inc., a revolutionary social entrepreneurship manufacturer that produces solar lighting equipment primarily for the developing world. He adds, "It wasn't just people Maren unified. It was departments and organizations. She was a politician." Patrick couldn't resist Maren's infectious enthusiasm and unyielding determination, whether it was to plan a crazy bike trek from Central Illinois to Argentina or to apply for funding for his start-up business. "I got so much out of this university, almost unfairly, because of Maren's abilities," he added. "I certainly would not be where I am today without her."

Classmate and friend Ben Barnes (BS, MS Mechanical Science and Engineering, 2006, 2009) shared a similar perspective on Maren, whom he had met in 2004 when they were both entering the world of the neophyte Engineers Without Borders campus chapter. During that year, Ben recalls, the organization flailed about and didn't have a lot of rules . . . until Maren took charge.

"'Forceful' was a learned behavior for Maren," Ben said. "But being driven came naturally to her. It was a passion, maybe almost a desperate passion. And she kept the idealism going for the whole organization because of her influence and maturity."

So who was this Maren Somers?

"Maren was always interested in improving the well-being of everyone she came in contact with from early childhood," recalled her mother, Nancy Hanna-Somers. "Engineering gave her a big range of tools. For Maren, school was something that wasn't easy for her, but she felt a responsibility to excel."

Growing up in a rural agricultural community, Maren looked to the Girl Scouts as an early opportunity to serve others while surrounding herself with strong women and girls, Nancy said. It was in the Scouts that she made her first big push for international experience by learning Spanish and earning money to travel to Europe with her troop. She also worked with her mother, a project manager for an environmental engineering firm, and was able to see the connection between engineering and water resources.

"I was actually surprised when she wanted to enroll in engineering," Nancy said. "But she was very drawn to the water issue, and saw a need for clean water."

It's no surprise, then, that Maren would be attracted to a new student organization dedicated to finding engineering solutions for international communities in need. The first trip she took with the group was to work on a biofuel project in India with Patrick and Ben, which derailed their ambitious plans to bike to Argentina.

"Traveling to India with Maren was an experience," Patrick said. "I don't think she'd traveled much, and I'd never even gone camping as a kid, then all of a sudden we're living in a village without electricity and running water. It was a completely mind-blowing experience."

Despite her inexperience in the developing world, Ben observed, Maren coalesced the students in the team with a unique adeptness, even though she wasn't the designated leader.

"Without Maren, that first implementation trip would have been a disaster," Ben said. "She hit harder, was more mature and more motivated than anyone else on the trip."

The combination of politician, unifier, and organizer moved Maren to the frontlines as new projects came online, and she became a force in the Nigeria water project that serves the village of Adu Achi, in the state of Enugu. While there, she connected with a younger student who was attracted to Engineers Without Borders by the promise of using her technological skills to develop engineering solutions that would serve communities in need.

Cheryl Weyant, a student in the department of Civil and Environmental Engineering, was "intrigued by technology, though I don't think I thought really deeply about it at the time. Maybe it was more the broad idea than the technology that lured me in." Past experiences volunteering abroad had discouraged Cheryl. She felt she hadn't made an impact other than to provide money, and had felt "in the way."

However, bolstered by a shot of Maren's energy, Cheryl tried international service once more, and this time she found something different.

"It was never about 'what's this work doing for me.' I got irked by people who did it as a résumé builder," Cheryl said. "With Engineers Without Borders, it really became about just caring for each other . . . it became a family of volunteers. And the push to excel came from teamwork."

Cheryl found herself gaining more than the social benefits of being a part of the volunteer family, too. Her skills in leadership, which continue to serve her as she pursues a PhD in environmental engineering under the tutelage of Tami Bond, blossomed in Engineers Without Borders.

Not that the new responsibility wasn't terrifying, Cheryl said. "It was odd to be called the leader by this group of people, because I never worked toward being a leader. I think it was being on the trips and being able to speak to what happened. But I now have a certain confidence in my capacity to lead, having been on a site, helping build (water storage) tanks, being the last responsible person making decisions."

For Cheryl, Engineers Without Borders meant going from being "a freshman sitting in the back of the class and never answering a question to sitting in the front and being more confident in speaking out, using my strengths, and asking broader, complex questions rather than just doing an equation on the board." Did the experience make her a better engineer? Cheryl's response: "I think I'm a better person."

Not every Engineers Without Borders volunteer was initially drawn to the organization by its emphasis on service. For at least one woman, technology was the initial draw. Hanting Wang had entered the University of Illinois in 2009 as a business student, but she knew before the end of her freshman year that the major wasn't a good fit. Though she explored Engineers Without Borders her sophomore year during an information night, it wasn't until her junior year, when she had transferred into civil and environmental engineering and was working on undergraduate research in biosand filtration, that she fully immersed herself in Engineers Without Borders project leadership.

"I came in as project lead halfway through a biosand filtration project in Guatemala, so (technical) implementation had already begun. But we had a lot of community issues and lots of distrust," said Hanting, who chose to participate in this particular project because of her research interests. "Through cleaning up that (social) mess, I learned a lot about paperwork, administration, bringing people together, and powering through it. It wasn't the technical skills that made a difference. It was the 'soft skills' that turned things around."

Perhaps because of that experience, Hanting considers herself "a little different" as an engineer. "I had developed soft skills in high school and through business, and I learned to be able to record things effectively and handle paperwork. And I can see big-picture things well. I saw recurring issues that weren't apparent to everyone else on the project." These skills serve her well now in her job as a consulting engineer for a civil-engineering firm in Chicago.

What Hanting calls "soft skills" may be less easily obtained in a classroom than they are in practice. More than a few women students have noted that applied engineering projects like those available through Engineers Without Borders allow them to merge their technical expertise with skills in negotiation, management, communication, and documentation.

"Engineers Without Borders really reinforced my confidence in my skills," Hanting said. "I met lots of environmental engineers in my same year, made friends, and found a comfort level. It really helped with my transition into engineering and created a very tight-knit group." That many of those friends were women in leadership continues to increase the comfort level for younger women students who explore the organization for the first time. Many first-time volunteers find inspiration when greeted by a leadership that is diverse and well represented across genders, ethnicity, and political leanings.

Diversity, coupled with a common passion for helping others, is what attracted the University's 2014 International Undergraduate Achievement Award winner, Nora Onstad, to join Engineers Without Borders when she arrived on campus in 2011.

"I felt really lost at the first meeting, but I saw the passion of others so I kept going," recalled Nora, who graduated with an Agricultural and Biological Engineering degree in 2015 and continues to pursue research at the graduate level. "Some professors, you can tell they love teaching because they're excited for an hour and some professors you can tell want to be done with the hour and get out. It was the same with students in Engineers Without Borders—the people I saw were waving their arms and very excited. You knew they wanted to be there."

Nora became so engaged with a water project in the west-central African nation of Cameroon that she spent five months living in the community of Ntisaw, researching the function and maintenance of the water system that her Engineers Without Borders team had designed and built. The experience provided her a tremendous capacity for accepting differences in attitude, culture, and expectations.

"In the isolation of Ntisaw, I found a community, but it could be difficult to (deal with) its rules sometimes. . . . For example, sexism is overt in Cameroon. But I tried to go with the flow, though it's hard not to place my own view of how things are done in the United States on this society."

Engineers Without Borders had already helped her to deal with the social mores revolving around men's and women's roles in society, since Nora found acceptance and camaraderie among her organization teammates in a way she didn't always experience through her classes.

"In some ways, Engineers Without Borders was really good because I got to see that other women succeed in engineering," Nora said. "Going into a

FIGURE 8.2 Maren Somers: "Peace on ya." Photo courtesy of Dr. Peter Somers.

discussion section when you're the only woman in the room is kind of disheartening. And when they had a luncheon for award winners, I looked around and it was all guys, except for me and the department secretaries.

But everyone in Engineers Without Borders is open and supportive, and having female professional mentors helped us to see that there are women who are actually doing the work we want to do. I want to have a role mentoring students in international projects someday."

The path these strong women walked at Engineers Without Borders was carved by powerful personalities like Maren Somers. But pioneering an organization dedicated to serving the world with passion and dedication can take its toll.

"Maren saw this effort as an obligation to the world, and she believed that if you can't help others, you shouldn't be able to sleep at night," Ben recalled. "She was a person who took things on herself."

That burden could be overwhelming at times, Patrick said, to the point that sleep began to elude her and her health suffered. Struggling with thyroid disease as well as clinical depression, Maren worked frantically both with Engineers Without Borders and in her graduate research in the Department of Agricultural and Biological Engineering, where she was studying ways to remediate water-source contamination due to cyanide. Cyanide contamination in water is common in Nigeria and results from processing cassava—a staple of the Nigerian diet.

"She felt an immense responsibility to finish the project in Nigeria, always deferring classes to keep all the balls in the air," Patrick said. "I knew she would stress sometimes but I had no inkling of how much she was struggling."

On June 20, 2010, Maren succumbed to the stress and responsibility, dying by suicide after she drank some of the cyanide solution that she was using in her graduate laboratory. Her family has established a scholarship that is given to a student in the Department of Civil and Environmental Engineering each year to honor her memory and the very important role she played in Engineers Without Borders. The commemoration of Maren's life and contributions to the developing world continues each year with a celebration at her parents' farm, and many of the organization's volunteers—both men and women—revel in recalling the no-holds-barred passion that she unleashed whenever the needs of a developing community were at stake.

"Her idealism was about making the world more just," Ben said. "And Engineers Without Borders was Maren's tool to do it."

Because of the women in this chapter, there is hope for cleaner air and cleaner water around the planet. There is also hope for the "trailing edge" of society that is more acutely affected by these problems. Tami Bond is optimistic about finding reasonable solutions that fit the context of each community. Her optimism comes partly from students: "They are smarter than we are," she says, "It's our job to nurture them and then stand aside."

Do Engineers Have to Engineer?

ANGELA S. WOLTERS

In the late 1920s, the notion of women using an engineering background for a nontechnical career was shared by Elsie Eaves in the book, *An Outline of Careers for Women: A Practical Guide to Achievement*. Elsie, who in 1927 had become the first female member of the American Society of Civil Engineers, was a practicing engineer-turned-manager of the Engineering News Reports (ENR), a weekly construction industry trade journal with a national audience.

In her chapter on civil engineering, Elsie noted: "Most engineering courses require four years of intensive work, with class schedules somewhat heavier than in other college departments . . . the training is systematic and analytical. An engineering education is a valuable training in fundamentals and an excellent background for a business career."[1] Throughout the history of engineering education, both men and women pursuing degrees in engineering have welcomed career opportunities that span from traditional to nontraditional engineering roles.

More recently, engineering degree programs have been touted as the "New Liberal Arts." In *Forbes* magazine in early 2016, Dinesh Paliwal—CEO of connected technologies company HARMAN—wrote, "Engineering has proven to be one of the most fruitful tracks of study in the job market, as the skills and training developed by an engineering program are far more versatile than many believed."[2] Engineers train in both creative and analytical thinking, a combination that prepares them to contribute in any number of ways to our society.

Surveys show that nearly 40 percent of those with bachelor's degrees in engineering work in nonengineering–related jobs. Engineers find careers in law, health care, management, business, entrepreneurship, government, policy, education, and the nonprofit sector, to name only a selection of possible fields.[3]

Many women engineers marry their technical training with additional passions as they pursue roles outside the box of standard engineering careers. Alumnae from the College of Engineering at Illinois have become accountants, actuaries, alderwomen, authors, business analysts, criminal investigators, dentists, doctors, economists, editors, entrepreneurs, financial advisors, lawyers, librarians, military officers, musicians, pastors, patent attorneys, pilots, real estate brokers, sales executives, school superintendents, surgeons, teachers, veterinarians, and writers. Three alumnae of note—Kit Gordon, Laurie Morvan, and Linda Reinhard—diverged from traditional engineering career paths but relied on their engineering training to reach their goals.

✳ Kathryn "Kit" Gordon (1961–)

(Adapted from an essay by Christine des Garennes[4])

As she tests water from creeks near her home in Los Altos Hills, Kit Gordon reflects on her successful career in semiconductors, entrepreneurial ventures, and volunteer work with environmental groups. Having spent nearly sixteen years as a successful engineer, Kit migrated away from engineering to pursue other passions and feels both contributed to her wonderful career working and living in the center of tech innovation.

As a newly minted chemical engineering graduate in 1983, Kit was drawn to Silicon Valley's young, informal, and innovative start-up culture. Her chemical engineering degree along with her determination and appreciation for hard work and teamwork enabled Kit to succeed in the semiconductor industry including obtaining numerous patents in the field. In college she learned "to work hard and to rebound after setbacks. You persevere and keep trying. The university taught me to take risks, something every engineer must do," Kit says.

Kit grew up the youngest of three children in Rockford, Illinois. Shortly after she was born, her father developed a brain tumor and became disabled. He passed away when she was twelve. Kit's mother, raised on a farm during the Depression, worked as a grade school teacher. When considering majors and careers, Kit wanted a career that would provide financial stability—and she knew chemical engineering could do that.

At Rockford West High School, she excelled in chemistry and math and planned to attend the University of Illinois. Even though she was a star student in chemistry, her high school did not offer honors or Advanced Placement classes. She took the required chemistry placement exam for chemical engineering majors and didn't pass. This was a setback, but she was determined to persevere. An advisor tried to steer her to another major but Kit was not deterred: "I knew what I wanted. I wanted to be a chemical engineer. I knew myself better than they knew me. I never wavered."

Her first year on campus was tough, but she knew once she started taking chemical engineering classes, she would be fine. She was right; she excelled. She enjoyed many classes—fluid dynamics, physical chemistry, organic chemistry, and all the chemical engineering classes. In her classes, she learned not only the value of hard work but also how to "divide and conquer" through teamwork.

Kit found community and inspiration by joining the university's chapter of the American Institute of Chemical Engineers. Once, she attended a meeting during which R. Byron Bird, Warren Stewart, and Edwin Lightfoot—chemical engineering legends and authors of the classic textbook *Transport Phenomena*—stood in the same room. She was in awe. During her senior year, she received the prestigious Daniel Eisele Award for outstanding student leadership in chemical engineering, which included a $400 award. She used the money to buy her first car—a 1969 AMC Rambler, fondly known as "The Scrambler."

That same year, Kit interviewed with a number of corporations. Most of her interviews consisted of formulaic questions asked in nondescript boardrooms. But her interview with Monolithic Memories (MMI), a Silicon Valley semiconductor firm, was different. No one wore a suit, most employees were young, and their enthusiasm was contagious. She toured the site, met employees, and learned about their projects. They asked her many technical semiconductor-related questions, and she found herself answering, "I don't know" to many of them. Given that her background was in chemical engineering and not electrical engineering, she recounted one exchange with her interviewer: "'Why are you asking me these technical questions outside my major?' I asked. He told me, 'A good engineer knows when to say I don't know. I want to hear you say I don't know. I don't want to hear you making things up.'" When she asked about standard promotion schedules, the interviewer laughed and said, "You can be president tomorrow if you can do the job."

Kit was sold on the culture of Silicon Valley. She moved to the Bay Area and was thrilled to see her name in the same phone book as Nobel laureate Linus Pauling—at Illinois, she had studied the Pauling Principle on electroneutrality. While working at MMI, she attended Santa Clara University and earned a master's degree in electrical engineering and later, an MBA.

Kit says that her time at the University of Illinois, particularly the group work and risk-taking, prepared her for the industry: "Working in semiconductors is a fast-changing, high-pressure, taking-lots-of-risks kind of a job. It involves groups of people doing complicated engineering to get a product out." She recalls a particularly formative experience in one of her early chemical engineering courses, when she scored 19 out of 100 on the first exam. The high score was 21. "That taught me you can fail and still be OK. It helped reinforce the grit I needed when working in an innovative industry. Failure is a great teacher."

Five years into her tenure at MMI, the company was purchased by Advanced Micro Devices. The three inventors of the PAL (programmable array logic) device used to implement logic functions in digital circuits spun off to form a new company, which ultimately became QuickLogic. Kit was the first person they hired. She invented the technology (amorphous silicon anti-fuse[5]) for the company's first products. After the company went public in 1999, she took time off of work, got married, and in 2002, gave birth to her daughter Isabel.

Around the same time, because of a growing concern about the types of chemicals her daughter was being exposed to, Kit began experimenting with natural skincare remedies. She and a friend established Botanic Organic, a plant-based skincare line. Kit has stepped back from day-to-day operations in that business but continues to make and use natural products for her family, mixing hydrosols, oils, and extracts. Currently she is interested in learning more about skincare products derived from bacteria.

Her interest in living sustainably and caring for the environment has its roots in Illinois. Her grandfather, an agricultural engineer, taught her at an early age about the interdependence of watershed health and human health. When her grandparents retired, they moved to a small farm along the Rock River. As a child, Kit spent many Sundays tromping around the farm, exploring the apple orchard, the woods, and the fields where farmers grew corn and oats. Her naturalist grandmother taught her to identify various birds and insects. And she recalls her grandfather telling her in the 1970s that her generation would need to fix the problems inflicted on the Ogallala aquifer and other groundwater basins. "He instilled in me a sense of environmental responsibility," she says.

In Santa Clara (aka Silicon) Valley, Kit has been heavily involved in a number of environmental organizations, including the Santa Clara Valley Water District, the public agency that manages water supply, flood protection, and stewardship of area streams. She served on several committees for the water district: the environment and water use committee, independent monitoring committee, and the One Water planning committee, which looks holistically at flood control, water supply, and environmental protection. She serves on the board

FIGURE 9.1 Kit Gordon: "Appreciating the importance and beauty of California's precious waterways." Photo courtesy of Tanya Radowicz Boillot.

of Grassroots Ecology, a watershed restoration group; Streamkeepers, which performs local water quality monitoring; and Permanente Reimagined, which is restoring steelhead trout to Permanente Creek in the Santa Clara Valley.

In May 2016, Kit returned to campus to share details of her story at the Chemical and Biomolecular Engineering convocation. She has also been planning for the future. She is considering taking classes at Stanford University, possibly in hydrology or ecosystems. She studied Ecology of a Meadow as an elective at Illinois, as well as Poetry, and continues to enjoy both subjects. While not sure what her next degree or project will be, but with her daughter off to high school and other family responsibilities reducing, Kit looks forward to her next adventure. Chemistry, she's sure, will lead the way.

✳ Laurie Morvan (1961–)

How does an electrical engineering graduate become the lead singer and guitarist of a blues band? Just ask Laurie (Watters) Morvan.

With an interest in knowing how things worked, Laurie chose electrical engineering as her college major. Her love for math and science prompted her high school counselor to nudge her to explore engineering. As a child, Laurie's stepfather, Bud King—an electrician and talented handyman—encouraged her and her siblings to use tools and learn how to fix things around the house. This training, paired with her natural interest, solidified her choice of major. Even today she spends time working with Bud to complete "Daddy-Daughter" home improvement projects on her home.

As a female student in electrical engineering in the early 1980s, Laurie was often the only girl in her classes. Occasionally, she encountered a male counterpart who didn't believe she should be there. During one upper-division circuits course, the male student sitting next to Laurie commented, "Shouldn't you be home dusting?" Without a comeback at the time, Laurie had a chance to rebuke him after the first exam. After writing all test scores on the board: 85, 63, 61, etc., the professor for the course handed back exams in order of highest to lowest grade. With the first exam handed back to Laurie—indicating that she received the highest score on the exam—she turned to the male student and said, "Dust this!"

Luckily, confronting this kind of sexism was not the norm for Laurie. She was a busy and involved student. In her sophomore year, in hopes of receiving some scholarship money, she tried out for the intercollegiate volleyball team. Having earned twelve varsity letters in high school—four of them in volleyball—Laurie not only made the team but secured a full-ride scholarship that supported her as she finished her degree. With an interest in the aerospace industry, Laurie also decided to pursue her pilot's licenses through the Institute of Aviation. Before graduating, she secured private, instrument, commercial, and multi-engine pilot's licenses.

Laurie's schedule was packed with engineering and flight classes sandwiched between early morning strength training and evening volleyball practices or games. She often missed class to travel to away matches and tournaments. To keep up with her course work, she would bring along her electrical engineering projects (e.g., circuit boards), packing them alongside her acoustic guitar.

Laurie was a flutist and drummer in the high school band, but it wasn't until her late teens that she discovered her instrument of choice. A friend introduced her to the acoustic guitar, and Laurie took to it so fast, she "wrote her first song after learning only three chords." At Illinois, during delayed flights to and from volleyball games, Laurie would break out the guitar and get the whole team singing.

Laurie's excellence in the classroom and on the court earned her the George Huff award in her senior year for her scholarly and athletic achievements. She also won the A. R. "Buck" Knight Award from the Electrical and Computer Engineering Department for her good scholarship and leadership in student activities outside of engineering. After graduation, she moved to Redondo Beach, California, and began working for an aerospace company. She also joined a rock-and-roll cover band as a rhythm guitarist and vocalist.

But Laurie wanted to be the lead guitarist. So she drew on her strong work ethic and practiced endless hours to perfect her technique. After three years of balancing engineering and music, Laurie quit her aerospace job to tour full time as lead guitarist and vocalist for a new band. She studied the recordings of the world's greatest rock-and-roll players, focusing on guitar-driven blues rock and southern rock. But Laurie's "whole life changed" when she was introduced to the music of Stevie Ray Vaughan. For Laurie, this introduction to Stevie's version of the blues was "like being turned loose in an infinite, beautiful new universe!"[6]

From there she found "what she was born to do" in creating her own version of red-hot blues rock. Next step was to record. In 1997, she did just that; she and her band, Backroad Shack, recorded *Out of the Woods*. But making an album isn't cheap. To fund her musical passion, Laurie decided to go back to graduate school to pursue a master's degree in applied mathematics from California State University. With her advanced degree, Laurie started teaching college math courses. "If I'm going to do something that takes me away from music," she said, "then it's going to be something that helps people."[7]

Currently, Laurie teaches math at Cypress College. Several of her courses are online, so she can balance her time between her teaching and the 50 to 80 days of touring her band does each year. In the fall of 2015, Laurie took a sabbatical and created an online trigonometry course, developing videos of all her lectures. Students loved being able to pace themselves as they learn and to return to lectures whenever they need to.

When asked how she balances both careers, Laurie says, "I love playing music and I love teaching math. It's hard to juggle both careers, and frankly, if I could make enough money playing music, I would have that as my only career. I feel blessed that if I have to have a 'day gig,' I get to do something as fulfilling and important as teaching. My students inspire and energize me, and I really enjoy being a part of their life's journey."

In 2004, Laurie officially renamed her band the Laurie Morvan Band and recorded *Find My Way Home*. Three years later, they released their third album, *Cures What Ails Ya*, which proved to be their turning point. *Guitar Player* magazine

FIGURE 9.2 Laurie Morvan rocks the stage at the 2013 Blues, Brews, and BBQ Festival in Champaign, Illinois. Reproduced by permission of *The News-Gazette* (2017). Permission does not imply endorsement.

featured Laurie and the album in their October 2007 issue, highlighting Laurie's songwriting talent—she has written the majority of songs on the albums. In February 2008, the band qualified for the finals of the International Blues Challenge for Best Self-Produced CD, only to return in 2010 and win the competition with their fourth CD, *Fire It Up*.

Successes abounded. In 2009, Laurie and her band were highlighted by the Big Ten Network in the production of "The Laurie Morvan Band: Nothing but the Blues," a television program based on personal interviews and footage from the Ellnora Guitar Festival, held in the Krannert Center for the Performing Arts on the University of Illinois campus. Champaign-Urbana's local newspaper, the *News-Gazette*, noted, "Morvan has all the soulfulness of Bonnie Raitt and the swaggering, muscular guitar tone of Stevie Ray Vaughan . . . a blistering, high-energy ending to another great guitar festival at Krannert."[8] In 2011, The

Laurie Morvan Band recorded their fifth album, *Breathe Deep*. Laurie says their five albums "cover the whole spectrum of happy to sad to mad and celebration and heartbreak. The human experience is this unending palette of inspiration."

In the future, Laurie wants to continue sharing her music. She'd like to create a recording that resonates with an even larger number of people. "Every minute I'm on stage is precious to me. It's a real honor to share your music with people. It's the sharing of the music and seeing people's reaction to it. It's huge. It's a huge, wonderful thing to do in life."[9]

Laurie's engineering degree taught her logic, order, and a methodical approach—skills she's applied to practicing and studying guitar and building her music career. Not only that, but she knows how to fix a whole bunch of her own equipment!

Both engineering and blues music are male-dominated fields, making Laurie something of an outlier. She's managed by taking slights in stride. "You have to choose," she says, "either you're going to be mad about that kind of stuff or you're just going to go. My job is just to work hard and when I get my opportunities and the light shines upon me I'm going to make sure the light is illuminating something worthy of people's attention. That's the path I've chosen."[10]

Laurie sums up these feelings in the second verse of her song, *Living in a Man's World*:[11]

Sometimes I've had to go twice as far
Sometimes I have to be twice as good
Just to get half as much respect baby
Half as much as I should
Cuz I'm a woman yeah, yeah, yeah
And I'm livin' in a man's world
Don't you feel sorry for me baby
Cuz I chose this life as a guitar playin' girl
I chose this life

✳ Reflections by Linda Reinhard

BS Electrical Engineering, 1986
Vice President, Strategic Relationships and Business Development,
Industry and Infrastructure at Informa

Does each student graduating with an engineering degree need to design or code? Based upon my own nontraditional experience, the answer is no; there are many paths to take with an engineering degree. Many business functions

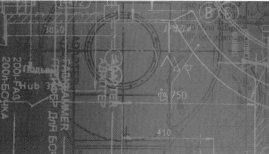

FIGURE 9.3 Linda Reinhard. Photo courtesy of Linda Reinhard.

require engineering knowledge. This knowledge is critical in making "big technology ideas" happen.

At age twelve, I became very interested in computers. My father had worked on the software for the Apollo Mission Simulator, which helped me see how computers could change the world. Luckily for me, my mother found a nearby computer camp at Western Illinois University in Macomb, Illinois, which I attended the summer before seventh grade. That's where my technology journey began. In high school, I continued to advance my computer skills by showing my teachers how to use computers, writing programs to keep statistics for various sports teams, and helping librarians move from paper to computer directories. I loved the work. It showed me firsthand how computers could change our daily lives.

During my valedictorian speech in 1982, I predicted that researchers might soon release a design for a self-controlled car in which the driver would program the destination and the car would take it from there. Ideas like this filled my mind and excited me. But I had to earn money to help pay for college, so I worked as a waitress at a local restaurant. There I learned that hard work, efficiency, and providing great customer service pay off.

I was very fortunate to have University of Illinois as my state school as it allowed me to study electrical engineering at one of the top programs in the world. As my study of computers deepened, I branched into radio frequency (RF) communications. The technical knowledge gave me critical competencies as my field entered the telecommunications boom.

I also participated in a liberal arts study abroad program. The European trip exposed me to new countries and cultures, giving me a perspective I couldn't gain in the classroom. Traveling abroad made the following semester challenging—it left me with eighteen hours of technical electrical engineering courses when I returned—but the experience was worth it.

Studying abroad and waiting tables led me to the conclusion that I wanted to mix my technical studies with people-oriented work. So in my junior year, I asked the head of the electrical engineering department if I could do a co-op program in technical sales. He said, "Hewlett Packard (HP) was just here asking for a co-op student in technical sales and I told them they would never find one!" I began working for HP right away in technical sales of computers and test equipment, spending the semester in Chicago. The following summer, I went to California to handle product marketing for HP. And after completing my degree in 1986, I decided to follow this career path, to contribute to the developments of marketable and world-changing technologies.

As a young sales representative for HP in 1987, I was tasked with a major challenge to unseat IBM's mainframe computers as the "go-to" computer system used in industry with HP's new Reduced Instruction Set Computing (RISC) computers. It was a gargantuan task, but the low cost of the RISC computers helped me sell the new systems to clients. Local area networks (LAN) were just beginning at this time, and my technical ability helped me explain and design a customer's network without having to involve an application engineer. Still, I wanted the additional understanding of the business world, so I entered Northwestern's Kellogg School of Management evening program to earn a Masters of Management.

After five years, I left HP to test my newfound strategy skills at Motorola's semiconductor sales group. In my role as Strategic Program Manager, I created the FAST Plan (Factory And Sales Team Plan) to integrate the Motorola equipment group's product needs into the semiconductor product roadmaps—the prediction of how the product will grow. After developing the plan, I wanted to help put it into action, so I moved back into sales. Within six months, I sold six custom semiconductors to the Motorola group responsible for the first wireless personal digital assistant (PDA), which enabled wireless email for the first time. It was also the first time I helped bring a big idea to market.

In 1996, I moved to Motorola's Cellular Infrastructure group based in Hong Kong. There I provided business development support to the local sales team and helped sell the new, high-capacity Code Division Multiple Access (CDMA) technology to China. Twenty years later, most cellular networks in China still use forms of CDMA technology, which optimizes for available bandwidth.

I came back to the United States in 1999 to serve as director of Product Management for a new venture between Cisco and Motorola, called Invisix. My team of product managers scoured the world for internet applications that we could make wireless. We integrated those applications into our enclosed wireless data network centers and gave hundreds of demonstrations of the new applications to major wireless carriers all around the world. The carriers began to deploy those wireless networks and applications in the early 2000s, and the wireless internet became a reality.

Meanwhile, in 2001 I had joined a start-up company called NAVTEQ. They were building navigation maps to be used in cars and to provide driving directions for customers such as AOL and Mapquest. As vice president for China Business with NAVTEQ, I managed the relationships with our China partner to collect the map data for China in the NAVTEQ format. I also worked closely with Nokia, the mobile phone market leader in China, to put the first navigation application on a mobile phone. And recently, NAVTEQ, now known as HERE, was purchased by a consortium of European Union (EU) automakers for the purposes of developing autonomous cars. So my vision of self-controlled cars that I articulated in my valedictorian speech may soon come to fruition.

Looking for opportunities to educate the energy industry, I joined Penton Media in 2014 as VP and Market Leader for Energy and Buildings, leading a group of technical editors and sales people who educate the professionals in each industry and also bring buyers and sellers together. In November of 2015, our team hosted an event in Chicago to bring together electric utilities, regulators, city government, building owners, and universities to help the industry chart a path to cleaner, smarter, more efficient cities of the future—another Big Technology Idea that is critical to our planet.

So, engineers do not have to engineer. Those with a propensity toward math and science with a love of problem solving, people, and other cultures, can find success and fulfillment in many areas. My path—selling great technology—requires many job functions: engineering, manufacturing, quality control, product management, product marketing, strategy, sales, and customer service. All of these roles benefit from a strong foundational understanding of how technology works.

Mentors and Mentoring

LAURA D. HAHN

In January 2010, President Barack Obama spoke at a reception in the White House for winners of the Presidential Awards for Excellence in Mathematics and Science:

> Every person in this room remembers a teacher or mentor that made a difference in their lives. Every person in this room remembers a moment in which an educator showed them something about the world—or something about themselves—that changed their lives. It could be a word of encouragement, a helping hand, or a lesson that sparked a question that ignited a passion and ultimately may have propelled a career. And innovators . . . scientists and engineers are made in those moments.[1]

Mentoring does make a difference. It affects the recruitment, retention, and advancement of women graduate students and faculty.[2] Focused, purposeful mentoring programs also benefit women undergraduate students[3] and demographically underrepresented students.[4] National initiatives such as the Million Women Mentors[5] program for girls and young women, international programs such as TechWomen[6] for women STEM leaders in Africa, Central Asia and the Middle East, and numerous locally based summer camps and after-school programs are on the increase.

Joi Mondisa knows; she's researched it. Joi earned a BS in General Engineering from the University of Illinois in 2001 and went on to earn a PhD in engineering education from Purdue. Focusing on mentoring for minority students in STEM, her work highlights the complex ways that mentors help their protégés—sharing personal experiences, instilling confidence and autonomy, and building shared knowledge and community are all profound investments in the success of others.[7] It is clear that good mentoring requires extraordinary commitment.

This chapter focuses on women in the College of Engineering who have supported, encouraged, and changed the lives of numerous students at Illinois. From 1946 to 1973, Grace Wilson nurtured women in the College by promoting community—through afternoon teas and the inception of the local chapter of the Society of Women Engineers. Geneva Belford provided counsel and academic assistance to countless students in computer science. And Susan Larson has worked to create new programs and activities to support women students in the College. What they have in common is their devotion to student success.

✳ Grace Wilson (1907–2005)

Grace Wilson was on the faculty of the General Engineering Department for twenty-seven years. When she began in 1946, there were 114 women enrolled in the College of Engineering (and 2,961 men). Apart from a dip in the early 1950s, the number of women students increased steadily throughout Grace's career and, at the time of her retirement in 1973, 73 women were enrolled, along with 2,964 men.[8] While Grace taught nearly all of these students, she spent much of her time supporting the women: their studies, their sense of belonging, and their career aspirations.

Grace was a "townie." She grew up in Urbana, Illinois, the daughter of Teresa Wilson, a homemaker, and Wilbur Wilson, a faculty member in the College of Engineering. Although she was undecided about a career path early on, she believed that she would never want to be a teacher. She discovered an interest in drawing through a high school physics class, where she had to draw a diagram of her own house and determine where to place heating units. Her affinity for drawing led her to enter the architecture program at the University of Illinois in 1925. She completed her degree in 1931, in the last class of architecture students to get degrees from the College of Engineering.

When Grace graduated from the University of Illinois, jobs were hard to come by. However, she managed to find a variety of jobs that involved design. She made brick scales for an architectural firm in Chicago, she designed bridges for the state of Michigan, and for a while she and a business partner operated a dressmaking shop in Chicago.

Although she swore she would never be a teacher, Grace came back to Champaign-Urbana in 1941 to teach mechanical and architectural drawing in the Champaign schools. Her role as a mentor and role model for young women began there—one of the high school girls heard that a woman would be teaching the class and decided to take it—as the only girl in the class. That student eventually went on to become an architect.[9]

In 1944, out of a sense of civic duty, Grace joined the Women's Army Corps' Air Transport Command and worked there as a draftsperson until 1945. She then came back to Champaign-Urbana and worked at the Small Homes Council.

In 1946, World War II veterans began enrolling at the University of Illinois, and the engineering classes were flooded. Associate Dean Harvey Jordan hired Grace as a teaching assistant for a night section of General Engineering 107, Geometry for Architects. She became an instructor the following semester with a full teaching load, and, by the time she retired in 1973, she had the title of full professor. In 1962, she coauthored the textbook, *Geometry for Architects* (Stipes Publishing Company). A second edition came out in 1975.

Grace would be the only woman faculty member in the College of Engineering until 1968, when Lorella Jones joined the physics faculty (see chapter 2, "Pioneers"). However, in 1962, Elisabeth "Tebby" Lyman came to Illinois as a research assistant in the Computer-based Education Research Lab (CERL), and Grace and Tebby joined forces in supporting the women students in the College.

One of Grace's most important contributions to the community of women engineering students at Illinois was undoubtedly the establishment of the Society of Women Engineers chapter. Four women students met at Grace's home on October 15, 1959, for the first formal planning meeting, and in 1964 SWE was officially established at Illinois. Three of the early presidents of SWE—and beneficiaries of Grace's mentoring—are friends to this day: Pamela Calvetti VanBlaricum, Elizabeth ("Betty") Richards, and Carolyn Primus.

Pam came to the University of Illinois as a freshman in aerospace engineering in 1968. She became active in numerous college-wide activities, and in 1972 she cochaired Engineering Open House. She was a James Scholar and a Knight of St. Pat, and later became the first woman to earn a PhD in Aerospace Engineering at Illinois. After graduation she worked for Effects Technology Incorporated (ETI) in Santa Barbara, California, testing the effects of underground nuclear weapons. She worked there full- and part-time, while raising two daughters, until she retired in 1996.

Betty Richards met Pam when she came to Illinois for a high school visit— Pam was assigned as her "big sister." The visit to Illinois was not her first; Betty had also spent a week at an NSF-sponsored summer camp for girls at Illinois. Both of these experiences reinforced her interest in engineering at Illinois. With the encouragement of her high school guidance counselor, she enrolled as a Ceramic Engineering major in 1969. She earned her BS in 1975, her MS in 1977, and a PhD in 1981. She spent her full career (twenty-seven years) at 3M.

Carolyn Primus also came to Illinois as a ceramic engineering major. In addition to being president of SWE, Carolyn was a Knight of St. Pat, Tau Beta Pi, and Top 100. She went on to earn degrees in Materials Science and Engineering at the University of California (MS Berkeley, 1975, and PhD Davis, 1980). She has worked in solar energy, naval materials, consumer goods, and medical devices. She holds thirteen patents and is the founder of Avalon Biomed Inc., a medical device manufacturer. In 2013, she and her husband established the Primus Engineering Scholarship Fund to support deserving underprivileged women engineering students at Illinois.

Pam met "Miss Wilson" on the first weekend she was on campus. Grace and Tebby Lyman were cohosting a luncheon for the 30 women in the College of Engineering. Pam remembers, "Everybody had to go around and talk. One girl said 'I'm just here because my dad wants me to be an engineer.' And another one said 'I don't know about engineering, but I really love math.' I think after a year or two, there were three of us left (out of seventeen freshmen)."

Pam describes how Grace and Tebby provided her and her peers with a sense of security and belonging: "They were there. They were like a safety net. If any of the women in engineering needed something, if you needed a ride somewhere (because none of us had cars), a shoulder to cry on, if you had some sort of problem, they would step in. We all trusted that."

Carolyn has a similar recollection: "When I arrived on campus, I received an invitation to a luncheon/tea and met Ms. Wilson and Dr. Lyman. For me it felt like having a life preserver thrown to me. At that time it was easy for a young woman to feel alone and isolated in engineering classes because we were so few. I was so grateful to meet these two women who each had created their own successful lives in science. Through them, and the other young women that were present, I felt reassured that I could survive the rigors of the classroom. The joy was to also have friends that had similar interests and goals. Their kindness gave me courage and a happier educational experience."

Betty recounts an example of the magnitude and impact of Grace's support: "In the summer before my junior year, Miss Wilson suggested I attend the national SWE conference in San Francisco. Somewhere, she found a fund to pay for my trip. My parents were reluctant to let me travel alone, but Miss

FIGURE 10.1 Pam VanBlaricum remarks: "I remember going up to the second floor of the Transportation Building to see Miss Wilson. . . . She worked standing up. Her door was always open and she was always perfectly attired. There was never a wrinkle in her blouses." Reproduced by permission of *The News-Gazette* (2017). Permission does not imply endorsement.

Wilson called them and reassured them that she'd help with the traveling and keep an eye on me in San Francisco. On the day of the flight, she drove the two-plus hours to my house in Gillespie and had lunch with my mother at our house. She left her car in our garage, and my parents drove us to the St. Louis airport. She and I spent four or five days at the convention before returning to St. Louis. The next year, I was offered a summer job at Bell Labs in Murray Hill, New Jersey. This time it wasn't difficult to convince my parents that I could fly to New Jersey by myself. If I hadn't gained travel experience with Miss Wilson, I might have missed a very exciting work experience, one that convinced me to go to grad school."

Grace was a beloved individual in the College of Engineering. She was made an Honorary Knight of St. Pat in 1969. And in 1973, the year Grace retired, the Illinois Society of Professional Engineers established the Grace Wilson Award for the outstanding woman student in the College of Engineering, based on GPA and activities. She passed away in 2005 at the age of 97.[10]

✳ Geneva Belford (1932–2014)

Laura Schmitt, Department of Bioengineering

Known for her encouraging nature and calm demeanor, Geneva Belford positively influenced the lives of hundreds of graduate students during her nearly fifty-year career at Illinois. As a computer science (CS) faculty member, she advised more than 140 graduate students, and toward the end of her career, she helped scores of additional students through her role as the graduate program coordinator in the CS Department.

According to CS Professor Roy Campbell, Geneva was unique because she regularly advised masters-level students while most faculty members preferred working with doctoral-level candidates. "She was always very welcoming and provided support for many students [interested] in distributed systems and networking," Campbell said. "She also took on a few atypical students, who didn't have a CS background."

One of those students was Tom Siebel (BA History, 1975), who entered the CS master's program in 1983 while working on an MBA at Illinois.[11] He earned his MS in CS in 1985 and went on to found Siebel Systems, a customer relationship management (CRM) firm that was later acquired by Oracle for $5.8 billion. In 2014, Siebel described his CS experience as a life-changing event. "Geneva was tough and practical, yet empathetic and generous with her time," Siebel said in a video tribute to his advisor when she won a 2012 CS department service award. "She encouraged exploration, experimentation and creativity. She was an important catalyst in my professional career."[12]

Paul Magelli (MS CS, 1986) appreciated the way Geneva pushed him and other students. "She took on talented MS students, but she made sure you set the bar high enough," said Magelli, founder of Chicago-based health analytics and data company, Apervita. "After reading an early draft of my thesis she said, 'Interesting summary, Paul. It's not enough. You can do better than this,'" he recalled. "I knew I could do better, too. In a very polite, professional way she essentially told me that she had higher standards and I had more capability."

In addition to her high standards for research, Geneva was a meticulous editor, providing students with valuable feedback on their journal papers and theses. "The thoroughness of her feedback was mind boggling," said Ajay Tirumala, who completed his MS with Geneva in 2001 and earned his doctorate with another CS professor in 2006.

After reading an early draft of Tirumala's thesis, Geneva suggested numerous improvements. "Her communication style was to encourage her students to do better and give their best, while she always put her students at ease."

Tirumala also admired Geneva for being a trailblazer in an era when the CS and math fields had very few women students and faculty. "She broke new pathways, earning her PhD in 1960 from a top-notch university," he said. "I use her as an example when I'm talking to my daughters. I'm grateful to have been associated with Professor Geneva. Her biggest impact is that she's a source of inspiration for all the young women who are trying to make it."

Born on May 18, 1932, in Washington, D.C., Geneva grew up in the Philadelphia area. She earned her bachelor's degree in math from the University of Pennsylvania in the early 1950s and married R. Linn Belford in 1954. The couple lived in California while Linn earned his PhD and then relocated to the University of Illinois when he took a faculty position in the chemistry department. In 1960, Geneva earned her PhD in mathematics, and after conducting research in the chemistry department for several years, she joined the mathematics department as an assistant professor.

In the early- and mid-1970s, Belford conducted research with the Center for Advanced Computation, including developing an experimental front-end interface for the Department of Defense's ARPANET network. She also helped develop file allocation studies and techniques for the World-Wide Military Command and Control System Intercomputer Network.

In 1977, Geneva became a faculty member in Computer Science, where she delved into distributed database research and quickly became one of the most popular advisors. She also established herself as an informal mentor to younger faculty members.

"She was always enthusiastic and collaborative about working with other people," said Campbell, who was a young assistant professor in 1977. "She made it easier for me to get into the distributed systems area because she had this broad view and memory. No matter how busy she was at the time, she'd welcome me into her office. She could pick out things I should look at, and she pointed me in the right direction."

Although she retired from the CS faculty in 2000, Geneva served as an advisor in the CS graduate student academic advising office, where she helped guide many students who were struggling with their classes, research, or personal issues. "I admired how she had a knack of looking at the big picture and could help students who may have been frustrated, scared, or timid," said Rhonda McElroy, a former CS academic advisor who worked with Geneva in the early-to mid-2000s. "She'd help move them to the next step in a positive way."

"She was a great mentor to me," added McElroy, who is currently director of Graduate Programs for the entire College of Engineering—a role that requires her to sometimes help students who aren't able to finish their PhD. "Geneva

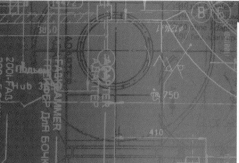

FIGURE 10.2 Geneva Belford faculty photo, ca. 1981. Photo courtesy of the University of Illinois Archives at Urbana Champaign.

had a great ability to talk a student through this difficult situation and help them find other options. I learned a lot of tips and approaches from watching her do this and I use some of those techniques today."

Geneva wrote about her mentoring views at the time she won the 2004 Campus Award for Excellence in Graduate Student Mentoring: "The task of a mentor is not just to get a student over technical difficulties, but also to provide encouragement and morale boosts as needed. I believe if you expect the best from students, they will generally meet the challenge."

Yong-Bin Kim (PhD CS, 2010), Geneva's last graduate student, received a great deal of that encouragement. A nontraditional student, Kim was married, had two young children, and was working full time in the last couple of years of completing his doctoral thesis. "We'd have weekly meetings to discuss my research," Kim said, "and sometimes I hadn't made much progress. A meeting like that with most advisors could be painful, but she was like mom—worried about me completing my program. She didn't want me to quit."

Today Kim is a senior engineer working at Caterpillar in the University of Illinois Research Park. "Were it not for her, I might not have finished my PhD degree," he said. "Now I'm working at a great company because I got my degree from one of the best schools. She was a great mentor."

During her Illinois career, Geneva's positive impact on students extended beyond the CS department to the entire campus. In the early 1980s, she served as an associate dean in the Graduate College. Later, she devoted much time to improving the university as a member of the faculty senate. She also shared her expertise as a member of campus committees that addressed challenges such as the implementation of online registration, educational technologies, and administrative data processing.

An outstanding educational leader, teacher, and mentor, Geneva earned several campus honors, including the 1986 College of Engineering Halliburton Educational Leadership Award, the 1991 Dad's Association Outstanding Faculty Award, the 2005 Graduate College Outstanding Mentor Award, the 2007 Mom's Association Medallion of Honor, and the 2012 CS @ ILLINOIS Distinguished Service Award.

Outside of CS, Geneva played flute for many years in the Parkland College Orchestra. She enjoyed painting, rock collecting, European history, and travel—she and her husband Linn, who were married for sixty years, especially enjoyed visiting Colorado each summer. She died March 4, 2014.[13]

Karen Hortig, executive director and CEO of the Society of Women Engineers, said, "The power of inspiration, mentorship, support and encouraging words from our community should not be underestimated."[14] We therefore conclude this chapter with a reflection from Assistant Dean Susan Larson, a mentor to women—students, faculty, and staff alike—from the College of Engineering who cross her path. Susan earned undergraduate degrees in physics and German at Washington University in St. Louis. She earned her MS and PhD in environmental engineering science at the California Institute of Technology. She came to the Department of Civil and Environmental Engineering at Illinois in 1988, with a research focus on the behavior of air pollutants. In addition to being an award-winning teacher and researcher, Susan is an active advisor and advocate for women students in the College.

✳ Reflections by Susan M. Larson

Assistant Dean, College of Engineering
Associate Professor, Civil and Environmental Engineering

Sometimes, I think that real mentoring is something one can only see in retrospect. Mentoring can be found in advice from someone you may have met only briefly, but somehow, what s/he says fits exactly with what you needed to hear, and you remember it always. Only further along do you recognize how this bit

FIGURE 10.3 Susan Larson (R) with student Lara Flasch. Photo courtesy of Angela Wolters.

of advice shaped you. Sometimes, looking back, you might see that mentoring arose out of a longer association: a friend, a classmate, or an advisor who took an interest in your professional development, encouraged you, help[ed] you build up a realistic confidence, and provided feedback. Looking at "retrospective mentoring" from the other side, whatever we do now to appropriately support and encourage classmates or colleagues may prove to be extremely valuable to them in the future. But perhaps this view is formed by my own experiences.

The first time I ever heard of a formal mentor role was in a job interview, and the question along the lines of "Does it bother you that you would not have a mentor in this job?" I think I was surprised twice. Once because I had never heard about formal mentoring, so the question implied it was something I had missed. The second surprise was the instinctive realization that not having a mentor would be a disadvantage for me in this job I so wanted. That was a long time ago, and I don't think the question would be posed today. In the last few decades, mentoring has become more of a recognized and planned part of education, training, and development. Mentors are often arranged for people new to an organization—first-year college students, new assistant professors, and recently hired employees. These arrangements—whether they involve multiple mentors or groups of protégées—can be especially effective in transmitting information on how things get done, what's expected, and who does what in the organization. However, formal mentoring programs may or may not succeed at the relational level. Just as putting two random roommates together may not result in a lifelong friendship, arranged mentoring programs may not end up with the supportive, encouraging mentoring that can shape a person or a career.

This deeper mentoring is relational. Its development can be encouraged, but it can't always be orchestrated. It's this deeper mentoring that people like Grace Wilson and Geneva Belford provided for their many students. It's the type of mentoring that Pam Calvetti VanBlaricum, Betty Richards, and Carolyn Primus gave to one another and to their fellow SWE members. Yes, it is important to design effective mentoring programs. But there will always be mentoring of great impact outside of formal mentoring programs. Alumni will look back and realize how a professor's lectures sparked an interest. Tenured professors will talk about how their thesis advisor's discussions impacted the direction of their research careers. Managers will remember aspects of the training they received when they were first hired. What I want to say is that no matter how good we get at mentoring programs, one person's own encouragement and support of others can have an untold impact.

Inspiring the Next Generation

ANGELA S. WOLTERS

In 1929, a new book called *An Outline of Careers for Women: A Practical Guide to Achievement* featured a chapter by Lillian Gilbreth. Lillian was known as the mother of Industrial Engineering. Lillian's chapter on industrial engineering featured words of encouragement: "The women in the profession have had only the most kindly and cooperative treatment from the men in the field, and have been given every opportunity for training, for work in the field, for advancement, and for participations in all activities of the profession. . . . This undoubtedly means that any woman fitted and willing to undertake the training and devote the time and effort to this work can succeed."[1] Lillian's shared experience seems nontypical given the history of women's lack of entry into engineering during the time. Nevertheless as an early pioneer, she shared these sentiments hoping to inspire women to pursue engineering.

Twenty-one years later, W. L. Everitt, dean of the College of Engineering at the University of Illinois, feared the nation was facing a critical shortage of approximately 25,000 new engineers per year. In December 1950, he called upon the engineering students at the University of Illinois to go home over winter break and share the news with high school students, high school teachers, and parents, stressing the "big present and future market for trained men and women in all branches of engineering."[2] Dean Everitt was a forward-thinking leader who saw women as the country's largest untapped resource for its future engineering workforce. His directed call encouraged women to study engineering.

Additional "calls" for women students continued throughout the 1950s, 1960s, and 1970s. Various institutions held events across the country to introduce women to engineering and inspire them to pursue engineering careers. While some private institutions kept their doors closed—the California Institute of Technology (Caltech), for example, didn't admit undergraduate women until 1970[3]—others like the University of Illinois tried a variety of methods of attracting women to consider further studies in engineering.

In 1962, for example, the College of Engineering crowned a Queen of the Engineering Open House (EOH), the University's annual student-run display of engineering projects. Before the event, the College asked high schools from around the state to nominate candidates for "princess" and "maid of honor" to be announced at the Knights of St. Patrick Ball—EOH's culminating event. The winners were awarded an all-expenses-paid trip to visit the EOH.[4] It was an effort to highlight the opportunities for girls and women who chose to study engineering and science. Later, in 1973, the University of Illinois sponsored the "Women in Engineering: It's Your Turn Now" conference, which allowed for sessions of informal conversations between high school juniors and seniors and college Society of Women Engineers (SWE) members.[5]

Over the years, many individual women engineers took the time to mentor and advise other women interested in the field of engineering. While they may go unnamed here, we recognize their efforts in the greater numbers of practicing women engineers today. These women worked hard to clear a path for themselves into the field of engineering and helped maintain that path as others made their way behind them. The two University of Illinois alumnae highlighted, Betty Lou Bailey and Sakshi Srivastava, stand for the many others who, with their passion and dedication, empowered other women to succeed in engineering and other STEM disciplines.

✳ Betty Lou Bailey (1929–2007)

"I signed up to take engineering because I thought I would like the work. It was not that I was . . . setting out to be a trailblazer."[6] Modest words. But regardless of her initial intentions, Betty Lou Bailey's successful career was both trailblazing and inspirational. The details she shared about her experiences resonated with many, making her an unintentional advocate for women engineers.

The daughter of a civil engineer, Betty Lou was the youngest of five children. Her oldest sister, Helen, urged her to study engineering. Helen had taught welding courses during World War II and noticed the growing opportunities for women in technical, male-dominated fields. In high school, Betty Lou worked as a file and mail clerk at the Chicago Bridge and Iron Company, an engineering firm. There she learned that "engineering would have less routine in the daily work than most other jobs."[7]

She decided to study engineering in college.[8] But she was hesitant to tell her father. She worried he would want her to enter a more traditionally feminine field. Growing up, he'd supplied Betty with "dolls instead of erector sets." She'd never learned to "[tinker] on a car."[9] As it turned out, Betty's father liked the idea. "He had seen women doing . . . unusual things during World War II,"[10] Betty Lou recalled, which may have made him more receptive to the idea of his daughter becoming an engineer.

With her father's approval, Betty Lou began her engineering studies at the University of Illinois in the fall of 1945. For Betty Lou, "the University of Illinois was synonymous with college. . . . That was where I wanted to go."[11] Betty Lou chose to major in mechanical engineering because of her interest in the mechanics principles of physics. Also, she held vivid memories of visiting her dad while he was working on a Tennessee Valley Authority (TVA) project in Knoxville, Tennessee. Betty Lou's mother took Betty and her sisters on a tour of a hosiery factory. "There were these big machines that were totally automated that were knitting the hosiery . . . it really was something that impressed me, and so I still remember it. . . . Little did they know that this little five-year-old girl was someday going to be a mechanical engineer."[12]

Even with significant interest in her chosen major, Betty was unsure that engineering would "work out" for her. During her first two years at the university, "once I walked out of the engineering classrooms and buildings. . . . I would not tell people I was an engineering student."[13] She felt isolated as a woman in the College of Engineering: "I had never known any women engineers directly or even through a mutual friend. . . . This, combined with a fear of not wanting to be different, meant that for the first two years of college, I was quite careful not to 'let it slip' that I was an engineering student."[14] The feeling of having to "hide" from her major was especially apparent when rushing the sorority that she would later join. During this time, Betty did not disclose her major; instead, she told them her interests were math and science even though the sorority had graduated a civil engineering student the year prior to Betty Lou's arrival. "I didn't want to make a point of it because I wasn't sure of it myself," she recalled.[15] It wasn't until her senior year that Betty Lou spoke openly about her major. The summer before, she had interned at Holley Carburetor in Detroit, Michigan, and had a great experience: "It built my own self-assurance to a considerable degree." She knew she had chosen a field of study with job opportunities that she would enjoy.

While on campus, Betty Lou spent most of her spare time outside the class-room working on nonengineering activities. She was involved in the Illinois yearbook and even served as associate editor her senior year. In addition to being active in her sorority, she served as an officer of Panhellenic during her senior year. In May of 1950, Betty Lou graduated in the top 3 percent from the College of Engineering as the only woman of the nearly 700 graduates in her class.

Prior to graduation, Betty Lou had secured her first job with General Electric (GE). Betty remembers attending a company information session and partici-pating in a twenty-minute interview, both on campus. During the interview, Betty asked whether the GE vice presidents were engineers or business majors. Engineers were the majority, and they encouraged Betty Lou as she "wanted to work for a company where engineers counted and were regarded as important."[16] The GE representative who interviewed Betty Lou reassured her that she would not be the only woman on staff. "Companies that are open to women tend to be companies that are open to new ideas," she reflected. "And if you are an engi-neer, male or female, you generally would prefer to work for a company where new ideas are welcome."[17]

GE made household appliances, a good career fit for Betty's early interests. During her training rotation program, Betty Lou was assigned to various sub-sectors of GE beyond appliances, including gyroscopic gun sights and steam turbines. She quickly realized that there were more opportunities for her if she wanted to take advantage of them.

Prior to her work rotation on the steam turbine in Schenectady, Betty Lou had been warned that the work was filthy because she'd be surrounded by oil. But she loved working at the plant anyway. The activity level was high, which "did a lot for me . . . physiologically. I then knew it didn't matter [that I was] a woman. And that does a lot for your bearing and how you behave afterwards." By the end of her rotation, Betty Lou felt comfortable working with her male peers: "I was one of them."[18]

After her rotations, Betty contributed as a testing, design, and system engi-neer in a variety of areas across GE's sectors, including jet engines, gas turbines, and space technology. She took her first permanent position with GE in the Cincinnati area working on jet engine design. She designed and patented a variable exhaust nozzle that allowed hot gas flow. Betty Lou considered this to be one of her most important engineering contributions.

In the 1960s, Betty Lou went back to school and gained her Master of Engi-neering degree through the Penn State Graduate Center. Also, she shifted to GE's Aerospace Division. This new assignment was quite different than her previous

FIGURE 11.1 Betty Lou Bailey, ca. 1962. Photo courtesy of the University of Illinois Archives at Urbana Champaign.

work. She led the designing, building, and testing of the adapter and payload separation system between a booster and the weather satellite Nimbus. The design, which was being conducted for NASA, included explosive devices that cut through a band to separate the components. The previous project manager had not been meeting the task schedule, and Betty Lou had been brought on to lead the project. But NASA resisted. They didn't want a woman as leader. Betty Lou's managers did not tell her this; instead, they let her present a three-hour technical project debriefing to a room of nearly fifty individuals from NASA and other engineering contractors. Afterward, NASA proceeded with the contract based upon Betty Lou's technical presentation. No one mentioned removing her as the lead on the project again.

Outside of her technical work at GE, Betty Lou became involved in the Society of Women Engineers (SWE), which was founded in 1950. During the summer of 1951, Betty Lou met one of SWE's founding members in Philadelphia, where Betty Lou had been assigned for three months to work on switch gears. Doris McNulty worked as a draftsman at GE during the day and went to engineering school at night. She introduced Betty Lou to other women engineers in the area. Betty Lou became a SWE member that summer.

Betty Lou saw great value in networking at SWE conferences. SWE members shared names of engineering managers that hired women. Finding out the names of the engineering managers was necessary so that women engineers could "bypass the human relations people," Betty Lou said. "Before the laws changed . . . [they] would automatically put your resume in the wastebasket." When SWE was established, the National Society of Professional Engineers conducted a survey and found that 65 percent of engineering employers would hire women engineers if they were available. While this was significantly less than 100 percent, Betty Lou emphasized that engineering was "such a large field that you don't need to have all the jobs open to you in order to have good job offers."[19]

Betty Lou also found SWE to be an important support network. SWE women shared similar experiences: "I soon learned in college that something that was hilarious that happened in engineering class, you come back to the sorority house, they don't even understand it . . . when you are with women engineers, you can cover anything that comes to mind."[20]

Betty Lou was very active on the professional guidance and education committees of SWE. She also helped with SWE's early outreach efforts. She enjoyed helping young women seek engineering as a career. In October of 1962, Betty Lou returned to the University of Illinois campus and gave a talk titled "Women Can Be Engineers" to a group of high school counselors.

She shared details from her own experience with the counselors so they could better counsel girls about engineering. Her thoughts are still relevant today. She opened with the following:

> I have come to speak on, "Women Can Be Engineers," and I'm undoubtedly supposed to be live evidence that this is indeed true . . .
>
> I believe there are some special problems in counseling girls about engineering and it is well worth understanding these problems.
>
> High school students are often going through a phase where they fear [being] different from others. This, coupled with a lack of knowledge about engineer-

ing, means that the majority of girls who would make good engineers probably don't even give it a thought . . . the young women of suitable abilities and traits need to be told about engineering as a career while they are still in high school. The "front door" is open for an engineering career, but the high school girls don't know it yet.

Betty Lou told of her own background and then went on to paint a picture of the opportunities that await new engineers:

What does await the new engineer when he or she graduates from college? As I see it, one of the greatest satisfactions of engineering work is the mental stimulation it provides. . . . Over and over again you are stretching mentally, reaching for answers. Since engineering jobs deal with practical problems and projects, these questions must be answered within a time limit. Consequently, engineering is characterized by finding a workable but not necessarily perfect solution to a given problem and then tackling the next quandary.

She spoke of other satisfactions, too, including having a sense of physical accomplishment and being on the forefront of science and engineering. She also commented on the need for engineers at the time:

In America, our engineering enrollment was 12% of our college students in 1950 and 9% in 1959. How are we to tackle the vast job of working out the practical application of a steadily growing body of scientific knowledge if our engineering enrollment is actually falling off slightly? . . . We in this country are dependent on the advice of counselors like yourselves who can recognize that we are in a scientific revolution and can tell the high school students that it behooves those who are qualified to participate fully by becoming engineers or scientists. Let me hasten to add that ability is needed and not sheer numbers. An obvious comment here is that girls should be just as thoroughly screened for talent as boys if we are to find a substantial means of increasing our engineers and scientists without sacrificing quality.

During her career, Betty Lou performed as a registered professional engineer in Ohio and New York and was also active in technical organizations, including the National Society of Professional Engineers, the Engineering Joint Council, and the American Society of Engineering Education. She was the first woman member in the Engineering Society of Cincinnati, an organization that focused on mentoring high school students. When Betty Lou joined, high school girls attended the outreach events but had no female role models. Betty Lou changed that by becoming a mentor for future women engineers. Her fame as a mentor is evidenced by a letter from a high school student written into the "Question-Air" column of the *Free Lance-Star* in 1967. This column was intended for people under the age of twenty to "ask a question of an outstanding personality that will be of interest to all teenagers." Mary Bennett of Somerset, Ohio, posed the following question:[21]

> I am in high school and just love math and science. I would like to be an engineer, but wonder if there are many careers open for a girl. Would you please ask Betty Lou Bailey, chairman, Society of Women Engineers, what subjects are necessary for me to take in order to have a career in any field of engineering that might be open to women?

The response from Betty Lou read:

> The young woman who has prepared herself for a career in engineering has a choice of a wide variety of professional jobs . . . although they are few, women engineers are well distributed throughout the engineering profession. . . . Most are in the fields of electrical, mechanical, civil, chemical, and aerospace engineers, just as the men are most likely to have jobs in design and research and development. . . . The job acceptance of women engineers is excellent. A survey of those graduating in 1966 showed that those who sought employment (instead of graduate work) averaged five job offers each. . . . Requirements for admission vary among colleges, but your interest in mathematics and science means you are probably already taking suitable courses in high school. . . . I hope you will find, as I have, that engineering is a career offering continuous mental stimulation, a feeling of practical accomplishment, and almost no routine work.

Betty Lou Bailey was an active SWE member her whole life, serving as part of the executive committee and becoming a fellow in 1985. In 1988, the Mechanical Engineering Department at the University of Illinois named her a distinguished alumna; she was the only woman to receive this award during its first forty-five years of existence.[22]

Outside of engineering, Betty Lou was committed to conservation and enjoyed the outdoors. An active member of the Adirondack Mountain Club, she used her engineering talents to help preserve nature by securing public access to free-flowing rivers.[23] Neil Woodworth, former Adirondack Mountain Club executive director and legislative counsel, remembers her contributions: "Betty Lou was often the only citizen advocate in the (Federal Energy Regulating Commission) hearing process. Betty Lou often managed to participate in 2 or 3 relicensing proceedings at a time while managing a demanding General Electric profession and her many outdoor recreational activities. . . . She skillfully knew the federal laws governing these FERC dam licenses and relicensing as well, or better, than the attorneys for the dam owners. She used her scientific analysis of the river characteristics, the dam's mode of operation and water levels to ensure both the health of the ecology of river life and the zest of the whitewater experience." Betty Lou also enjoyed traveling, canoeing, and hiking, and even trekked the entire Appalachian Mountain Trail by walking it in segments over a few years.[24] Her passion for engineering combined with her love for the outdoors added up to a life devoted to improving conditions for all.

Another alumna of the College of Engineering—Joan Mitchell—has empowered other women to succeed in engineering. Her mentorship efforts, including the authorship of a book on this topic, are highlighted in chapter 4, "Relentless Innovators."

It was with this same passion that Betty Lou took on the counseling of future women engineers. Most women engineers are supportive of other women pursuing degrees and careers in engineering and other STEM majors, but few truly take on the role of counselor and advocate. Betty Lou Bailey did exactly that at a time when it was drastically needed.

Sakshi Srivastava graduated from Illinois in 2015 with a bachelor's degree in electrical engineering. She's now pursuing her doctoral degree in the same department. During her time as an undergraduate, she led an effort to secure a women in engineering statue on campus to inspire current students and future generations of women students of engineering. Here is the story of her journey and the inspiration she found and shared at Illinois.

✴ Reflections by Sakshi Srivastava

BS Electrical Engineering, 2015; PhD Candidate

There are days so important that they set the course of the rest of your life. One such day was when I understood Faraday's law. Changing magnetic flux induces an electromotive force (EMF) in a closed circuit. The induced EMF is proportional to negative time rate of change of magnetic flux. Faraday's law made me appreciate the controlled behavior of nature. The world seeks equilibrium. That day, I decided to pursue engineering physics as my major.

The importance of good education was instilled in me since childhood. I was a high school sophomore when I came across the list of the best engineering schools in the world. It included Massachusetts Institute of Technology, University of California-Berkeley, Stanford University, University of Illinois at Urbana-Champaign, and University of Texas-Austin, all schools I had never heard of before. On doing further research about these schools, I was in awe of the facilities and resources that were available for students. From world-renowned faculty to undergraduate research opportunities, each of these institutions is a Mecca for aspiring engineers.

I wanted an engineering physics degree but I also knew that the odds were not in my favor to get admitted. I come from Allahabad, India, where there was no culture of going abroad for undergraduate studies. Consequently, I didn't know anything about Scholastic Aptitude Tests (SAT), Test of English as a Foreign Language (ToEFL), letters of recommendation, and the application packet for college admissions. Students with more resources and private tutors than I had were also applying. They had access to people who could guide them to seek experiences that would look impressive on a college application; I had the internet. I checked the application requirements for different schools. What immediately caught my attention was the statement of purpose. It was a place to tell my story. In India, students are ranked and allotted engineering college based on entrance exams alone. In the United States, universities seemed to be interested in me, a person, a student. Embracing the idea of not reducing human abilities to number or rank, I became all the more excited to be a part of this education system.

The days leading to the exam consisted of writing essays and taking practice exams. We had to go to New Delhi because that was one of the few exam centers in India. I thought I was prepared, but my score suggested otherwise. I felt dejected and unsure of what to do next. Taking the SAT was expensive in India. I didn't know whether my parents would encourage me to sit for it again. I approached my mother and asked for advice. "What are the repercussions of taking the SAT again?" she asked. "Will it hurt more than help?" I said, "I don't think so." And she responded, "Well, then you should take the exam again."

I realized that my biggest hurdle in taking the SAT was bearing the pressure for four hours, so I trained myself to overcome it. I went through enough practice exams that I started seeing patterns in the respective sections. I was able to categorize questions of one kind together and work through them. I took the SATs again, and got the score that I was proud to include in my college application. Satisfied with my performance, I took the subject tests and aced them as well. My best friend helped me prepare for the ToEFL; her extensive English skills came in handy. In retrospect, this experience taught me the importance of failure in one's growth. We fear failure because of the stigma attached to it. The moment we start learning from them, our failures, our mistakes, transcend into life lessons. And it is through these life lessons we grow wiser. Our struggles make us stronger.

I kept this dream of mine a secret. Only a handful of people knew because I didn't want to get the hopes of my teachers and friends too high. When the cat did get out of the bag, people asked questions like, "Why isn't your drama over yet?" and "Do you really think you will be able to get in?" I coped with this negative feedback by focusing on the work at hand. What mattered to me was the support from my family and friends. Whenever I was scared of the outcome, I would tell myself that even if I fail, I would fail on my terms. It is better to have failed than have spent a lifetime wondering, "What if?"

I was admitted to only one of the approximately ten schools I applied to—the Engineering Physics program at the University of Illinois at Urbana-Champaign. I was not offered any financial assistance. Consequently, I decided to decline the offer and save my family the emotional and financial burden. My parents are both medical doctors. They have spent their lives providing affordable health-care to the public. With a rupee-to-dollar ratio of approximately 45 in 2011, my education was estimated to cost about 10 million rupees. Three days after I had conveyed my decision to my parents, my father called me to his study. He had his and my mother's tax returns with him and he explained that if he sent both their salaries to Illinois for the next four years, we would be able to afford college. He told me that he was not able to attend one of the best medical schools in India because my grandfather could not pay for it; he wanted me to have the opportunity that he did not receive. In that moment, he made me believe that I deserved to know what the future at Illinois had in store for me. For everything that my parents sacrificed, I am forever grateful, for their support and how it has inspired me.

On August 15th, 2011, I finally made it to Illinois, ready to make the most of the acceptance letter that came my way. With my family in India and no phone or map with me, my arrival was challenging. I missed my bus from O'Hare to Champaign, somehow managed to get on the next one but then got lost on campus while carrying over 100 pounds in suitcases and my backpack. I finally found my way to my residence hall by asking students on the street for directions and getting a ride from a stranger. It was an eventful day. I attended the Women in Engineering (WIE) Freshman Orientation among many other student orientations. As school began, I enjoyed my time in classes but got regularly homesick on Friday nights. There was one other problem. I didn't feel as passionate about my major as I perceived fellow freshmen did about theirs.

Uncertainty is frightening. I felt guilty about investing my parents' money in something I wasn't sure of. I went to the advisors in the College of Engineering and discussed my options. I learned that I could change my major. I knew that I'd either be an electrical engineer or a mechanical engineer. Since I liked electromagnetics more than mechanics in high school, I petitioned to take the introductory course in electrical and computer engineering (ECE). It was in Spring 2012 when I fell in love with school; I belonged in the ECE department. I officially changed my major in Fall 2012. Throughout the process, I was anxious, but the advisors in the ECE and Physics departments not only answered my questions, but also made me feel that it was okay to change majors. Through this I learned to not be afraid to ask questions and for help. In times of need, a helping hand can make all the difference.

I believe that a sense of belonging plays a pivotal role in making us happy and content. The ECE department made me feel welcomed, but by being a woman engineer, part of that sense of belonging was compromised. At Illinois and at many schools around the United States, on average there are four men studying engineering for every woman doing the same.[25] I came from an all girls' school in India, so for me, this was a drastic demographic change. I wanted to understand the cause behind the skewed numbers. During my freshman year, I wrote a research paper for my rhetoric class on the underrepresentation of women in STEM (Science, Technology, Engineering, and Mathematics). I found out that women tend not to pursue STEM fields because they don't see themselves represented in the community. There are not enough women CEOs, professors, and leaders to act as role models for young girls: "The perception of female scientists occupying a shadowy wasteland on the peripheries of their departments is neither encouraging for potential role models nor their protégés. Low visibility of female scientists may discourage PhD students from consolidating an academic career in their discipline."[26] According to ASEE (American Society for Engineering Education), women earned only 20 percent of the total number of engineering bachelor's degrees and as low as 12.5 percent of bachelor's degrees in electrical engineering in 2015.[27]

In my junior year, I was appointed to be a resident advisor (RA). All RAs were required to take a leadership course. This course, coupled with the RA experience, helped me gain a deeper understanding of privilege, diversity, and inclusion. I realized that ignorance and innate biases can lead anyone into quick judgments. After interacting with students from diverse backgrounds, I learned to acknowledge my privileges and also my minority status, especially as an international student and a woman engineer. Based on the newly acquired knowledge and responsibility towards the community, I came up with an idea. It was to erect a woman engineer statue on the engineering campus, in a way to show our commitment to promote women in engineering. I took the idea to the College of Engineering and was advised to start an online petition.

The petition went live in Fall 2013. Meanwhile, I partnered with a student senator to seek support from the Illinois Student Senate. As days went on, students started to voice their opinions. While most community members embraced the idea, some lashed out at it. It was challenging to read through some of the negative comments on the social media. Part of me was enraged and another part of me was disappointed. Then a friend mentioned that the number of likes on these social media posts was significantly greater than the discouraging comments. So, I buckled down and read through the comments. I prepared answers to the questions I was asked by the critics and used their opinions to make my argument more robust.

On March 10, 2014, I addressed the Illinois Academic senate, explaining the need and support for a statue to represent women in engineering on the Illinois campus. A statue would convey a message to current students on campus that they belong in engineering, as well as inspire young girls who visit campus to picture themselves represented within the engineering community. The statue was approved.

By the beginning of 2015, the formations of a student committee was underway. It consisted of representatives from multiple engineering student organizations. This committee helped make pivotal decisions regarding the statue including the location and attire and the quote on the plaque. The project was student-run from the beginning to the end. Angela Wolters, director of Women in Engineering, and I served as the liaisons between the student and staff committees. That spring, Texas Instruments made a generous contribution that provided the funds needed to build the statue. The project received a lot of support from the College and the ECE department, and for that, I am thankful.

On April 28, 2017, the statue dedication ceremony was held. It was an incredible experience to live my dream. I shared the following sentiments when I addressed the audience during the dedication ceremony:

> But my job is far from done. Until a girl sitting in a remote village halfway across the world can aspire to be an engineer, we have to keep spreading awareness and empowering women. So, if you ever end up on the engineering campus of University of Illinois at Urbana-Champaign, don't forget to take a picture of the statue and post it on social media with #womanengineerstatue. Feel free to tweet it to SWE, IEEE, HeforShe, or to any news channel. Anyone who can listen and help us spread the word, can help us in getting our job done faster.

The day Texas Instruments announced their support for the statue in March 2015, Dean Cangellaris told me that I should never forget that it takes one person to change the world. He mentioned that going from 0 to 1 is the hardest step and after 1, the count gets easier. Following his words, this is my advice for anyone aspiring to be an engineer. Do not be afraid to be the person who changes the world. Follow your dreams. If you are genuinely passionate about a cause and

FIGURE 11.2 Sakshi Srivastava. Photo courtesy of the Electrical and Computer Engineering Department, University of Illinois.

work with a group of people who believe in it, you can bring about change. It is a matter of finding the problem you want to solve, and what you are willing to struggle for. Your strength is not in never failing, but in never giving up.

My journey from India to Illinois and the statue project have been an integral part of my life. These experiences taught me to not stop dreaming big. The first step towards achieving the impossible is conceiving that you can do the impossible. I had the honor of being a speaker at a TEDxUIUC in 2016. I shared the lessons I have learned by trying to do something new for the first time, coming to a foreign land for education and trying to initiate a social change. After the talk, it was very rewarding to hear from the members of the audience that they related to the message or found inspiration in it. I remember one person mentioned that he would share the talk with his sister who was attending engineering school then, to give her extra motivation to finish her program. As much as I cherished being a speaker, I enjoyed being an audience member even more. I went in with the goal to inspire others; I came out inspired myself. I learned that we can find inspiration—in ourselves, in our friends, our acquaintances. We just have to ask, "Do you mind sharing your story with me?"

The stories of Betty Lou and Sakshi are only two of the numerous stories of University of Illinois alumnae who have shared their passion for empowering other women to succeed in engineering and other STEM disciplines. Beyond the work of alumnae, various program offices on campus have also undertaken these efforts. Collaboration between the College's Women in Engineering (WIE) program and the student section of SWE led to the establishment of outreach events to empower women to study engineering.

One such event—Introduce a Girl to Engineering Day—was first implemented at the University of Illinois in 2014 through collaboration between Angie Wolters, director of Women in Engineering, and Emily Matijevich and Janna Eaves, codirectors of Outreach for the University of Illinois section of SWE. A national effort since 2001, Introduce a Girl to Engineering Day—spearheaded by the National Society of Professional Engineers (NSPE), IBM, and National Engineers Week Foundation—is a national movement to show girls and young women that engineering is a creative and collaborative field that is changing the world. Nationally, the "Girl Day" event introduces more than a million girls to engineering. Through efforts across the country and around the world, work continues to inspire the next generation of women engineers.

Foundations for the Future

JULIA STACKLER,
Department of Mechanical Science
and Engineering

We've already seen how throughout the late nineteenth and early twentieth centuries, when opportunities for women in engineering were sparse, many women overcame discrimination and societal obstacles to achieve greatness—paving the way to advance the field for women everywhere.

Hedy Lamarr, for example, the Austrian film actress, invented remote-controlled communications for the U.S. military during WWII that serve as the basis for Bluetooth and Wifi connections today.[1] In 1883, Emily Roebling took over the building of the Brooklyn Bridge after her husband became paralyzed.[2] She was the first female field engineer and technical leader of the project. Stephanie Louise Kwolek, a chemist at DuPont for more than forty years, discovered liquid crystalline polymers that became Kevlar—the stronger-than-steel product used in bulletproof vests, airplane fuselages, and fiber optic cables.[3] In 1903, Mary Anderson created the windshield wiper, an invention we now can't live without.[4]

But women like these didn't always make headlines. We can look back at their historical achievements and easily forget that women in every society in the world are still excluded from opportunities in science, technology, engineering, and math—and the impact is real. We can no longer reject the creative contributions of half the population; the grand challenges of the world demand that all perspectives are heard and that young women continue to enter STEM fields.

Now, as much as a century later, our country and our university are making great strides. Engineering is still a male-dominated discipline, but as of 2016, women comprised about 16 percent of tenured and tenure-track faculty nationally, representing a gain of about 4 percent since 2006. Of the approximately 35,400 tenured and tenure-track engineering faculty across the United States in 2016, the University of Illinois at Urbana-Champaign has the largest number (451), as well as the third largest number of women tenured/tenure-track faculty (76).[5]

In the last six years alone, the Department of Mechanical Science and Engineering (MechSE) at Illinois welcomed *ten* new women professors—and women now comprise 20 percent of the department's total faculty, ahead of the national curve. "Out of several hundred applicants, we interviewed many excellent candidates—with an even split of men and women—to be part of MechSE. Many of the women candidates excelled during the interview to the point that we believed it was imperative to have them on our faculty. They were the best of the best. And they are already making an impact here in both education and research," said Professor Andrew Alleyne, who was chair of the department's Faculty Recruiting Committee from 2012 to 2014.

The following profiles of MechSE's 2014 academic year class of five women faculty—Assistant Professors Alison Dunn, Yuhang Hu, Shelby Hutchens, Mariana Kersh, and Kelly Stephani—offer a glimpse into the breadth of ideas and innovation that can result from opening our doors wide to *all* talent.

✳ Alison Dunn (1981–)

Like the other women engineers featured in this chapter, Alison was influenced by the experiences of others in her family—especially her grandfathers, one of whom was a construction engineer and the other an aerospace engineer. "I liked taking things apart and building stuff so I chose mechanical engineering. And the more I went through the program, the more I liked it. I also had this specific vision from my grandfathers of what an engineer's career looks like, and I think that influenced my choices."

Alison's undergraduate experience at the University of Florida was a bit different than the average student. She got involved in research early on, giving her invaluable experience and a uniquely smooth transition into graduate school. Women made up about 25 percent—two out of eight students—of her research group at Florida, and the opportunities to work with other women outside her lab were slim. But, she said, "It wasn't something I put much thought into until someone brought it up to me. When I started taking engineering classes during my junior and senior years, I really thought that most majors were like that, where it was, say, 10 percent female—it never even occurred to me, because I was just in it, head down, moving forward, and there was never a group with all women in a class. I was naive and it wasn't on my radar."

After earning her master's degree in mechanical engineering, Alison chose the path less traveled: she and her husband decided to take a two-year break to work with the U.S. Peace Corps. They were stationed in China, teaching English. Back in the United States, the experience led to revelations for both of them. While her husband realized he didn't want to teach again, Alison did and knew a PhD in engineering was in her future. Her independent but positive experiences in research and teaching ultimately led her to a career in academia.

When she interviewed with MechSE, she had one young child and another on the way, so finding a position she could consider to be permanent was important. "At Illinois, we found really great people, a small town like where we came from, and the depth of research—and the friendly but very sharp people concentrated in this small little place was very practical."

While there weren't specific women in the department who served as inspiration during her interview process, Alison said she met several who shared similar experiences in education and in life. "I am really inspired by people, male or female, who can put things into practice. In an academic setting, it is very easy to just absorb, but our job is to turn it around and use it to do really good research and teaching and to build collaborations. I want to be in a place where research can be supported. There are women here in materials science, physics—who are all over the map—who can succeed. It's not just our department, but even being in the same place as these people and knowing they can succeed is pretty inspiring."

In her lab, the Materials Tribology Laboratory, Alison's research centers on what she calls "nontraditional tribology." *Tribology* is the study of surfaces moving against each other, and she is most interested in natural tribological systems, specifically, any part of the body that has a sliding interface—like knees, hips, and eyes—and therefore has the potential for disease associated with damage. Reducing friction, particularly of an implant in the body, could extend the life of implants and cause less damage or rejection.

In the field of engineering, it's easy to assume that all women engineers seek out a well-formed community of female faculty. Alison points out that perhaps because of the low numbers of females in the field, and in mechanical engineering in particular, some women may enter the field more suited to

FIGURE 12.1 (L-R) Mariana Kersh, Shelby Hutchens, Yuhang Hu, and Alison Dunn, 2017. Photo courtesy of Julia Stackler.

work in an environment that's male-dominated. The more established women engineers who thrived within—and despite of—the male-dominated system perhaps don't necessarily see all the opportunities that come with more women entering the faculty. "There may still be an undercurrent way back in our minds that we should work with men because in the past they, as a population, have been more successful as engineers. Women might be fighting this internal bias even when we are with each other. I think there are multiple tiers of change that have to happen and I don't know if we're out of the water in any sense. And I don't know the best way to counteract it besides working with people who do good work—female and male. We have some really successful women engineers here who have followed that strategy and have been here for years. Hopefully this new class of faculty will do that too. Maybe every woman who makes it has some exceptional quality that somehow overcomes traditional barriers, but I don't know what it is."

ALISON'S ADVICE FOR FUTURE GENERATIONS OF WOMEN ENGINEERS:

"Do your homework *well*. If you're female, and your homework is the clearest and you have the right answers, then the grades and everything will be there. They'll know clearly that you're the better candidate. If you decide to follow the status quo and do sloppy work, you give people the opportunity to not give it to you. You need to let people see your potential and offer you a job. Spend time on getting things right. Also, work hard and talk to people. Students who have the most trouble are the ones who don't talk to people, don't know how to collaborate, and assume that they can do everything themselves. That's something I have had to learn because I am an introvert and I would rather do everything myself. As a woman who wants to enter a technical field of any kind, do really good work and try to relate to people the best way you can without manipulation of any kind. They might not think you're as cool. But when you apply for the same job as someone else, you'll look a lot better."

✳ Yuhang Hu (1982–)

"The meaning of my name, Yuhang, in Chinese is 'exploring the outer space,' like an astronaut. Because of this, I thought I wanted to learn how to make things fly. I chose engineering because I was really good at math and physics in high school, and I think engineering is more of a handy skill, and it's something I really like." Her father, grandfather, and grandmother were all teachers, and they influenced her choice of career. "It was very straightforward for me. I wanted to become a good teacher."

As an undergraduate student studying engineering mechanics at Shanghai Jiao Tong University, Yuhang worked hard to meet the expectations behind her name, exploring the unknown and challenging herself. "I was surrounded by very competitive people, and most students were hard working. But I had a bigger goal of going to the United States to pursue a higher degree, so I worked very hard," she said.

Although the early days of her career in China and Singapore were not propelled by the assistance of strong mentors or much diversity, by the time Yuhang arrived at Harvard to pursue a PhD and then again at Illinois for her first tenure-track position, she was surrounded by a diverse scientific community where interdisciplinary collaborations were promoted.

"Before I came to the University of Illinois, I couldn't help noticing the work of the good scholars here. I feel very honored to work with them, and I hope I can do similar quality of work."

Yuhang studies pitcher plants—carnivorous plants with modified leaves that form a deep, prey-trapping cavity—to analyze certain bio-inspired systems and soft materials. The phenomenon with these particular plants is that their surface becomes very slippery when wet, causing ants, for instance, to slide down inside and get digested. However, the pitcher plant's surface is slippery only occasionally. When dry, that same surface acts as a sort of adhesive, allowing insects to walk easily on it and access the nectar from the plant. This varying surface property is what led her to develop a new material system with optical properties and wettability that can be continuously tuned by mechanical stimuli.

"From my basic training in solid mechanics and from my previous research on poroelastic properties of hydrogel, which is also a type of solid-liquid composite soft material, I know that if the material is stretched, it will change the pore pressure inside, causing the liquid to flow," she said. Based on this central idea, she and other researchers successfully developed a synthetic material that has continuously adjustable characteristics. "Such a material is made of a liquid film supported by a nanoporous elastic substrate. As the substrate deforms, the liquid flows within the pores, causing the smooth and defect-free surface to roughen through a continuous range of topographies. We show that a graded mechanical stimulus can be directly translated into finely tuned, dynamic adjustments of optical transparency and wettability. In particular, we demonstrate simultaneous control of the film's transparency and its ability to continuously manipulate various low-surface-tension droplets from free-sliding to pinned. This strategy should make possible the rational design of tunable, multifunctional adaptive materials for a broad range of applications," she said.

In addition to pitcher plants, Yuhang analyzes other biological systems that could potentially be used to create new materials and devices with multifunctionalities. She is interested in materials or systems composed of soft solid and liquid, and she studies the fundamental mechanics behind the nature systems that affect adaptability and efficiency. She also develops robust mechanical testing techniques to characterize these delicate materials.

She said she sees a lot of potential in her work for future study in biology and bioengineering—on the one hand, using the bio-inspired material and device to tailor the morphology of cells and tissues, while on the other hand, applying the mechanical testing technique to softer materials that she developed during her PhD work to characterize engineered biological materials.

Yuhang said that among the new, younger faculty at Illinois, there is a lot of opportunity to interact and organize activities or initiate research together. The sense of community she feels here, however, rests mostly on shared interests, which actually help inspire interdisciplinary research. The diversity and size of the university virtually guarantee that no matter the topic, there is someone at Illinois working on it.

"The university environment in the U.S. is also really good at encouraging minority faculty," she adds, "and there are strong Chinese communities here that support us."

YUHANG'S ADVICE FOR FUTURE GENERATIONS OF WOMEN ENGINEERS:

"From my past experience, I think perseverance is very important. My career path was always straightforward, but certainly we meet some difficulties and frustrations. Early on, my English was not very good, so I decided to spend two years building a strong foundation, language-wise and research-wise, so I could build a good background and whole package—giving me a better chance of getting to the U.S. I had to slow down things at that time. So just be persistent, and know that you can get there."

✳ Shelby Hutchens (1981–)

For Shelby, science runs in the family.

Her brother went into chemistry while her sister studied bioengineering. Her father, an engineer, and her mother, a pharmacist and accountant, instilled in their children the importance of a thorough education. Her parents' outlook undoubtedly influenced her decision to go to graduate school.

"I think that there are a lot of people like me—women in engineering—with fathers in engineering who thought their daughters could do anything, which is great. But one of the fundamental reasons I wanted to go to graduate school was that a lot of what I learned in undergrad just seemed like an introduction to a bunch of topics, and that was very dissatisfying to me."

While earning her undergraduate degree from Oklahoma State University, Shelby also played on the women's basketball team, which gave her a unique perspective. On the court, she was surrounded by many positive female influences, but in the classroom, she was one of only a few women in her chemical engineering program.

Still, at the time she didn't think in terms of gender. "Later on people would ask about my experience as a female in engineering," she said. "But people don't typically ask men, 'How is your experience as a man in science?'"

During her master's work at the California Institute of Technology, Shelby was part of a statistical thermodynamics research group, conducting analytical calculations on a theory of nucleation—charged macro-molecule solutions.

For her PhD research, she wanted to turn her focus to biological materials, so she transferred to a research group with Julia Greer, whose expertise is in both mechanics and materials sciences. Shelby's experience working in the varied groups uncovered her true interest in soft materials.

At Illinois, her research continues to focus on characterizing soft materials—primarily polymers—at small scales and studying soft materials fracture. "I'm interested in developing a new class of materials that uses plants as inspiration. In plants, you have a closed-cell cellular solid that you can load with salt to establish osmotic pressure gradients across the material. When this material is placed in water, it will deploy in some way based on how the structure of the solid is architected. By tuning the material properties and the geometry, we can create a set of materials that can provide an inhomogeneous stress state, or an inhomogeneous strain at the surface."

Shelby believes these properties could be useful in the future as implantable "soft casts" to help heal soft tissue that's been traumatized. "If soft materials modeled after plant tissue could be designed in a knowledgeable way, they might be able to help collagen fibers align during healing, and lead to better recovery of the tissue's original function."

What attracted Shelby to Illinois' MechSE department were the abundant opportunities for collaboration and the shared research facilities and equipment. She feels grateful to be part of such a large and diverse cohort of women in engineering, because it gives her the luxury of having many female perspectives—a rarity for previous generations of women engineers.

SHELBY'S ADVICE FOR FUTURE GENERATIONS OF WOMEN ENGINEERS:

"Try to figure out what kinds of problems are actually addressed in the various areas of engineering. Don't just be swayed by the primary tracks of study or by whatever is available. Look at other institutions besides your own as well to gain some perspective on what interests you."

✳ Mariana Kersh (1976–)

Mariana Kersh took the indirect path to becoming an engineer. But it seemed obvious from the time she was a little girl that engineering would be in her future. "My parents would buy furniture and I always wanted to assemble it by myself. And my dad would say, 'You're going to be an engineer!' He was right. I just took the long way."

Mariana first earned a bachelor's degree in English and worked in human resources before coming to the realization that she had wanted to work in medicine all along. But she wasn't truly committed to the long road to becoming a doctor. Fortunately, while enrolled in her second undergraduate program, she had several advisors who pointed her in the direction of bioengineering and biomechanics. "That is how I learned about engineering in medicine and that it was engineers who made the surgical tools and the implant devices, not the doctors."

Ten years later, Mariana had bachelor's and master's degrees in mechanical engineering and a PhD in materials science, all from the University of Wisconsin-Madison. "If it wasn't for Jia-Ling Lin from UW's tutoring program in the College of Engineering, I doubt I would be an engineer. I don't know for sure, but I know it would have been a much longer road to get here, and it would have taken me longer to find my space."

Later, during her postdoctoral position at the University of Melbourne in Australia, when Mariana was considering an offer from an orthopedics company, yet another advisor encouraged her to refocus on her ultimate goal of becoming an engineer. "He said to me directly, 'You're good enough, you do good work, and you have what it takes.' It was really powerful for someone to say that to me directly."

With this advice, Mariana took deliberate and specific steps to make herself a competitive candidate for faculty positions. She extended her postdoc by a year, which allowed her to finish more papers to make her résumé more robust. She sought feedback from her mentors about how she could further engage in teaching. And she set her sights on finding a university that would offer her the best resources to reach her research goals and achieve tenure. With a recognized and supported research area in biomechanics, along with exceptional imaging facilities, especially at the Beckman Institute, MechSE at Illinois fulfilled these criteria.

Mariana said that being part of such a large cohort of new faculty—with so many other women—offered a safety net in the midst of a demanding period, when there was so much to learn. Now several years later, the group still meets for monthly dinners, where they offer support and share their experiences applying for grants.

Mariana has also found a mentor in MechSE Professor Elizabeth Hsiao-Wecksler, a successful researcher who studies locomotion biomechanics and assistive device development. "She has been acting as my mentor since I started here. We have developed a level of trust with each other and even though we applied for the same funding at one point, she didn't see it as a competition. She felt strongly that we should both go for it without the worry that one of us would limit the other."

Although Mariana has formed professional relationships with both men and women during her studies and now at Illinois, she doesn't see them through a gendered perspective. "I don't think about the gender thing too much. It's really more about personality and seeking out people you like working with. Because it's mostly men in engineering, all of my mentors with the exception of my PhD advisor were men. But gender is not intentionally part of my thought process." Even so, she has also been involved in helping to mentor the next generation of female graduate students in engineering through regular meetings over coffee with women from MechSE's GraMS group (Graduate MechSE Students), offering advice, for example, about balancing work and family life. "I try really hard not to take it too seriously, even though we are all these high-achieving alpha people. I still like having fun and being a bit silly, and it sort of bums me out that you really can't do that with students in the classroom. This is a tricky thing to balance, and I think for women it's much different. We're already fighting that battle [to be taken seriously]."

In her Tissue Biomechanics Lab, Mariana's research focuses on the structural and mechanical properties of musculoskeletal tissues, using clinical-level medical images and the finite-element method (using mathematical approximations) to better understand and develop treatments for bone and joint diseases. She received funding to conduct research on the assessment of knee bone and cartilage after subchondroplasty procedures—specifically to examine whether subchondroplasty helps to inhibit further degeneration of the cartilage, which leads to osteoarthritis. By understanding the long-term effects of the procedure on bone and cartilage strength, and by understanding where the problems of osteoarthritis start, she said her research could present other potential treatments for subchondral bone defects.

"It's a dream to have such an amazing job. I am studying exactly what I want to study and I get to work with the surgeons and patients and orthopedic manufacturers and bring it all together. This job is my back door to medicine. I'm not a surgeon but we work together as partners."

MARIANA'S ADVICE FOR FUTURE GENERATIONS OF WOMEN ENGINEERS:

"Ask a lot of questions, and find people who are willing to listen. We're not born with any of this knowledge, but there's this idea that we should just know it, or that it should be organic or some inner stirring will push us to our path. I think being brave enough to ask those questions is a big part of it. We live in a culture, especially as engineers—and I see it in the classroom too—that if you don't understand things right away, there's a bit of shame. But asking questions is hard to do. And I don't always do it well, or I'll ask too late. You need to

FIGURE 12.2 (L-R) Kelly Stephani, Mariana Kersh, Shelby Hutchens, and Alison Dunn, 2014. Photo courtesy of UI News Bureau.

surround yourself with people who help you be confident and feel really free. They say to surround yourself with smart people. And I think that's true, but I've realized they are smart because they are always asking questions of the people around them."

✳ Kelly Stephani (1983–)

Throughout her childhood in eastern Wisconsin, Kelly Stephani always had an affinity for math and science and early on decided that the pursuit of an engineering degree seemed practical and would offer relatively flexible career options. "Biology and chemistry were interesting too, but math, physics, and calculus were also very challenging, which I appreciated. I wanted to explore things I wasn't necessarily comfortable with. So that definitely attracted me to engineering and the field tapped into the aspects of math and science that I really liked," she said.

Kelly was also fortunate to have parents who were supportive of her gaining hands-on and mechanical experiences as a child, encouraging her and her brother to understand how to do things like work on a car and build model rockets.

At the University of Minnesota, where Kelly was an aerospace engineering undergraduate, she had professors who inspired her to pursue graduate studies—to want to dive deeper into engineering. "I took the time to explore different options during my undergraduate years. I had internships and I was a research assistant, and doing these things made me realize that if I didn't go to grad school, within five years I was going to be bored."

Seven years later, she left the University of Texas at Austin with master's and doctoral degrees in aerospace engineering. Her PhD work was motivated by the well-known Space Shuttle *Columbia* disaster in 2003 in which there was a malfunction of the thermal protection system. Kelly is still inspired by understanding catastrophic issues in space travel and preventing future ones. "A lot of the work my group does is really dedicated to predicting how well an engineering system can perform in extreme environments. Everything we do somehow contributes to our ability to design safer systems—better hypersonic cruise vehicles for the Air Force, for example, or more effective thermal protection systems for NASA's shuttle reentry."

The research Kelly conducts in her Computational Kinetics Group focuses on modeling the conditions pertaining to non-equilibrium effects in gas dynamics. For a system with a lot of processes—like chemical reactions or thermal non-equilibrium—and changes taking place, understanding their chemistry and transport processes is an important component in devising her modeling techniques. She uses a special set of kinetic methods to devise computational tools that simulate the individual atoms and molecules in a gas at a particular level—tracking their motions and interactions. Kelly also researches gas-surface interactions, analyzing how gas particles interact with a solid surface in the presence of complicating factors such as chemical reactions.

Despite having an educational background grounded fully in aerospace engineering, Kelly believes there are many common features between the aerospace and mechanical engineering communities, and the wide range of applications of her work are far-reaching in both fields: high-temperature thermodynamic studies for space travel and atmospheric flight, micro- and nanotechnology, operation of micro-electro-mechanical systems (MEMS) devices, materials processing, plasma processing, and combustion and explosives.

"I'm in a field that has been pretty well established but now we're trying to push it into problems that become more complex and involve more complex processes. That's when it really starts to become challenging because the basis used to define the original set of equations may not hold, so we have to devise new ones," she said.

Kelly's perspective is similar to Mariana's: she doesn't necessarily experience a stronger connection with her female colleagues than with her community of scientists in general. "The way I see it is that we're all professors here, and we're all very interested in the problems we're trying to solve. In terms of a community of women, there are certainly activities we'll organize sometimes—we'll go to lunch or have coffee. But I find myself connecting with people I'm working with or I want to do research with. And I think that's a good thing! I appreciate that we're all considered equals and it's a very friendly and energetic research environment. It's pretty easy to find your place and start working with some really outstanding researchers."

Ultimately, Kelly wants to be known as *the* person in her field—to be at the head of the field, and have some input into where the field needs to go next, helping to propel the research community forward. She also strives to one day direct a large research center to collaborate with top researchers and tackle some of the big engineering problems. "I believe it's really up to scientists and engineers to motivate new areas of study that can help avoid future catastrophic events."

KELLY'S ADVICE FOR FUTURE GENERATIONS OF WOMEN ENGINEERS:

"Get as much experience as you can and do what you love. There's no need to enter a field just because you should. You need to feel like it's something you want to do and you can't see anything else fitting. I think if you enjoy the work but you're not fully in, it can be very challenging to want to stay in the field—because things don't come easy right away. And explore your options. Be fully informed. Get out and work internships, work with a researcher, and get those experiences early on. If you can't see yourself continuing along those same lines for ten years at least, you might want to consider other options."

These individuals, along with the other women faculty in the College, stand poised to make a tremendous impact—not only through their ideas and innovations, but through their inspiration of generations of students to come.

Closing

The stories of and reflections by the women in this book speak for themselves.

But *Women and Ideas in Engineering: Twelve Stories from Illinois* does not end here. There are still untold stories. And each day, there are new stories to tell.

Join us online for the continuing story of women engineers at Illinois: go.illinois.edu/WomenEngineers.

Women Graduates of the College of Engineering, 1867–1917

This list provides a summary of the first eleven women graduates of the College of Engineering. The list is a best faith effort to identify women graduates through review of records held by the University of Illinois Archives, including Board of Trustees Reports, Illio yearbooks, and alumni biographical files. Other sources including alumni databases were used to identify individuals.

In an effort to properly recognize woman graduates, this list will also be maintained/updated to correct any missing or inaccurate records on the following website: go.illinois.edu/WomenEngineers.

Name	Degree	Major	Year of Graduation	Additional Known Degrees
Mary Louisa Page	bachelor of science	Architecture	1879	
Mel Dora (Ice) Stritesky	bachelor of science	Architecture	1897	
Ethel Ricker	bachelor of science	Architecture	1904	
Alice Hartzel (Clark) Myers	bachelor of science	Architecture	1905	

Name	Degree	Major	Year of Graduation	Additional Known Degrees
Edith Leonard	bachelor of science	Architecture	1906	
Helen Jane (Van Meter) Alyea	bachelor of science	Architecture	1906	
Arselia Bessie (Martin) Swisher	bachelor of science	Architecture (Architectural Decoration)	1909	MS 1910
Louise Josephine Pellens	bachelor of science	Architecture (Architectural Decoration)	1909	
Nellie Nancy Hornor	bachelor of science	Physics	1912	MS 1913
Dorthea Marion Clayberg	bachelor of science	Architecture	1916	
Marion I. Manley	bachelor of science	Architecture	1917	

Women Graduates of the College of Engineering, 1918–1967

This list provides a summary of the women graduates of the College of Engineering between 1918 and 1967. The list is a best faith effort to identify women graduates through review of records held by the University of Illinois Archives, including Board of Trustees Reports, Illio yearbooks, and alumni biographical files. Other sources including alumni databases were used to identify individuals.

In an effort to properly recognize woman graduates, this list will also be maintained/updated to correct any missing or inaccurate records on the following website: go.illinois.edu/WomenEngineers.

Name	Degree	Major	Year of Graduation	Additional Known Degrees
Beryl Love Bristow	bachelor of science	Physics	1918	MS 1919
Margaret Kate Dawson	master of science	Physics	1919	BA Cornell College, 1917

Name	Degree	Major	Year of Graduation	Additional Known Degrees
Grace Greenwood Spencer	bachelor of science	Chemical Engineering	1922	First woman to receive Chemical Engineering degree from the University of Illinois*
Eleanor Frances (Seiler) Wittman	doctorate	Physics	1922	AB, AM University of Denver, 1913, 1914; AM 1916
Alberta Raffl	bachelor of science	Architecture	1923	
Geneva Fleming	bachelor of science	Architecture	1923	
Fay Morrow (Harris) Spencer	bachelor of science	Architecture	1923	
Carolyn Lindquist	bachelor of science	Ceramic Engineering	1923	
Mary Theye Worthen	bachelor of science	Architecture	1926	
Kathleen Boldt (Dean) Krueger	bachelor of science	Ceramics	1927	
Margaret Rothman	bachelor of science	Architectural Engineering	1929	
Minnie Isobel Mette	bachelor of science	Architectural Engineering	1930	
Dorothy Veronica Mary Held	bachelor of science	Architecture	1930	
Grace Wilson	bachelor of science	Architecture	1931	MS 1943
Mary Thelma Miller	bachelor of science	Civil Engineering	1933	

Name	Degree	Major	Year of Graduation	Additional Known Degrees
Sister Maria C. Voss	master of science	Physics	1938	AB, AM St. Mary's College, 1920, 1921
Marianne Ruth Freundlich	master of science	Physics	1944	BS Queens College, 1943
Lois Hume Windhorst	bachelor of science	Electrical Engineering	1945	
Anne Katherine (Lindberg) Williams	bachelor of science	Mechanical Engineering	1945	
Rosalyn (Sussman) Yalow	doctorate	Physics	1945	AB Hunter College, 1941
Harriet Reese Wisely	master of science	Ceramics	1946	
Barbara Lee Crawford Johnson	bachelor of science	General Engineering	1946	
Gladys Ida (Heinlein) Fuller	master of arts	Physics	1946	AB Hunter College, 1945
Mae Anna (Driscoll) Silbergeld	master of science	Physics	1946	BS University of Dayton, 1945
Marion A. Taggert Sherk	master of arts	Physics	1947	AB Smith College, 1943
Elisabeth Ethel (Farber) Lanzl	master of arts	Physics	1947	
Carol Elaine (Kousnetz) Sterkin	master of science	Physics	1947	BS University of Chicago, 1944
Barbara Emily Jordan	bachelor of science	Agricultural Engineering	1948	
Esther W. Miller Tuttle	doctorate	Ceramic Engineering	1948	BS, MS Alfred University, 1942, 1943

Name	Degree	Major	Year of Graduation	Additional Known Degrees
Barbara Lou (Schmidt) Hornkohl	bachelor of science	Civil Engineering	1948	
Tressa Eileen (Kimble) Mabry	bachelor of science	Electrical Engineering	1948	
Margaret Ellen (O'Donnell) Moran	bachelor of science	Metallurgical Engineering	1948	
Grace Charbonnet Moulton	master of science	Physics	1948	AB Tulane University, 1944
Helen Kava Zabinsky	bachelor of science	Aeronautical Engineering	1950	
Betty Lou Bailey	bachelor of science	Mechanical Engineering	1950	
Consuelo Wright (Minnich) Hauser	bachelor of science	Civil Engineering	1951	
Joan Lee (Hessler) Battuello	bachelor of science	Electrical Engineering	1951	
Barbara Gertrude (Engert) Ryder	bachelor of science	Mechanical Engineering	1951	
Ora Morlier Kromhout	master of science	Physics	1951	
Marcia Alden (Doolittle) Chipman	master of science	Physics	1951	AB Mount Holyoke College, 1949
Freda Friedman Salzman	master of science	Physics	1951	AB Brooklyn College, 1949; PhD University of Illinois, 1953
Marilyn Hoopes McKeon	master of science	Physics	1951	AB Wellesley University, 1947

Name	Degree	Major	Year of Graduation	Additional Known Degrees
Nancy Brazell Brooks	master of science	Civil Engineering	1953	B. Arch. Alabama Polytechnic Institute, 1950
Lily Hannah Seshu	master of science	Electrical Engineering	1953	BS, MS University of Travancore, 1944, 1946
Hanka Wanda Sobczak Chryssafopoulos	master of science	Civil Engineering	1954	C.E., E.E., M.E. University of Rio Grande deSul, 1951, 1953; PhD 1964
Gloria Esther Winkel DeWit	bachelor of science	Physics	1954	MS 1955; PhD 1959
Jane van Winkle Morgan	master of science	Physics	1954	AB Mount Holyoke College, 1952; PhD 1962
Chung Sul (Youn) Kim	bachelor of science	Chemical Engineering	1955	
Dolly Love (Marsh) Gurrola	master of science	Civil Engineering	1955	B. Arch. Alabama Polytechnic Institute, 1952
Mary Elizabeth Cunningham-Lusby	master of science	Physics	1955	AB Mount Holyoke College, 1953
Mary Koehler Fung	bachelor of science	Chemical Engineering	1956	
Eleanor (Owen) Heath	master of science	Civil Engineering	1956	BS Louisiana Polytechnic Institute, 1955

Name	Degree	Major	Year of Graduation	Additional Known Degrees
Shirley Janet Smith	master of science	Mechanical Engineering	1956	BS 1950
Marcia MacDonald Neugebauer	master of science	Physics	1956	AB Cornell University, 1954
Luda Aldona (Luinys) Kuprenas	bachelor of science	Chemical Engineering	1957	
Janine Anele Soliunas	bachelor of science	Electrical Engineering	1957	
Babu Lal Maheshwari	master of science	Mechanical Engineering	1957	
Ruth Fleischmann Weiner	master of science	Physics	1957	BS 1956
Meilute Ona (Tapulionis) Kusak	bachelor of science	Chemical Engineering	1958	
Patricia Helen (Hogeveen) Idu	bachelor of science	Civil Engineering	1959	
Lynn Dana McNames	bachelor of science	Electrical Engineering	1959	
Olga Marie (Ercegovac) Mosillo	bachelor of science	General Engineering	1959	
Roberta Ann Di Novi	master of science	Physics	1959	BS Fordham University, 1957
Leone Phyllis Murphy	bachelor of science	Ceramic Engineering	1960	
Doris Margaret (Maroney) Krumwiede	master of science	Ceramic Engineering	1960	BS Alfred University, 1958
Alice Anne (Summerbell) Culbert	master of science	Physics	1960	AB Oberlin College, 1958

Name	Degree	Major	Year of Graduation	Additional Known Degrees
Judith Catherine Rosenbaum Franz	master of science	Physics	1961	BA Cornell University, 1959; MS 1961; PhD, 1965
Nancy Marie McCumber O'Fallon	master of science	Nuclear Engineering	1961	BS St. Louis University, 1960
Martha Carolyn Schultz	master of science	Electrical Engineering	1962	BS University of Missouri School of Mines and Metallurgy, 1961
Dalia Terese Stake	bachelor of science	General Engineering	1962	
Franca Tagliabue Kuchnir	master of science	Physics	1962	Bacharel, University of Sao Paulo, 1958; PhD University of Illinois, 1966
Jane Hsia Shen	master of science	Civil Engineering	1963	
Sherrill Lee Wohlwend	bachelor of science	Mechanical Engineering	1963	
Patricia Mae Bunn	master of science	Metallurgical Engineering	1963	AB St. Joseph College (Connecticut), 1960
Theresa Shu-Yi Wang	master of science	Physics	1963	BS National Taiwan University, 1960

Name	Degree	Major	Year of Graduation	Additional Known Degrees
Joanne Elisabeth Pratt	master of science	Physics	1963	AB Smith College, 1961
Gerlina Louise Keltner	bachelor of science	Aeronautical and Astronautical Engineering	1964	
Adrienne Lois Lucchesi	bachelor of science	Ceramic Engineering	1964	
Judith Lavinia Hamilton	master of science	Civil Engineering	1964	AB Smith College, 1961
Kathryn Ann Miller	bachelor of science	Metallurgical Engineering	1964	
Sonia Reva Kronick Metropole	doctorate	Theoretical and Applied Mechanics	1964	RS, MS University of Michigan, 1959, 1960
Carole Ann Johnson	bachelor of science	Aeronautical and Astronautical Engineering	1965	
Janice Salmonson Helgason	master of science	Civil Engineering	1965	BS Buena Vista College, 1963
Francoise R.F. Proix	doctorate	Chemical Physics	1966	Diploma, Ecole Superieure de Physique et Chimie Industrielles (France), 1961; MS 1963
Carolyn Kay Lester	bachelor of science	General Engineering	1966	
Lois Backer Roberts	bachelor of science	General Engineering	1966	

Name	Degree	Major	Year of Graduation	Additional Known Degrees
Elizabeth Hicks Gardner	bachelor of science	Mechanical Engineering	1966	
Patricia Shirley Sora	bachelor of science	Mechanical Engineering	1966	
Nancy Marie McCumber (O'Fallon) Dowdy	doctorate	Physics	1966	BS St. Louis University, 1960; MS 1961
Franca Tagliabue Kuchnir	doctorate	Physics	1966	Bacharel, University of Sao Paulo (Brazil), 1958; MS 1962
Mary Elizabeth Jacobs	master of science	Physics	1966	AB Trinity College, 1964
Allison Dean Russell	master of science	Physics	1966	BS MIT, 1964; PhD University of Illinois, 1971
Donna Ortgiesen Jaske	bachelor of science	Aeronautical and Astronautical Engineering	1967	
Nancy Ann Barkley	bachelor of science	Aeronautical and Astronautical Engineering	1967	
Gerlina Louise Keltner	master of science	Aeronautical and Astronautical Engineering	1967	BS 1964
Janice Marie Ricker	bachelor of science	Ceramic Engineering	1967	
Sandra Collins Levey	bachelor of science	Civil Engineering	1967	

Name	Degree	Major	Year of Graduation	Additional Known Degrees
Nancy Lee Mary Mallonee	master of science	Civil Engineering	1967	
Stacy Lee Sachs	bachelor of science	Mechanical Engineering	1967	
Priscilla Jane Colwell	master of science	Physics	1967	AB Emmanuel College, 1965; PhD University of Illinois, 1971
Wendy Nan Torrance (Potter) Padgett	master of science	Physics	1967	AB Occidental College, 1965; PhD University of Illinois, 1971

* Chemical Engineering was and is a major in the College of Liberal Arts and Sciences at the University of Illinois.

NOTES

Preface

1. Akhila Satish, "Paving the Way for More Women in STEM," *Forbes Elevate*, September 4, 2014, http://www.forbes.com/sites/ellevate/2014/09/04/paving-the-way-for-more-women-in-stem/#266815a61823. Heather R. Huhman, "STEM Fields and the Gender Gap: Where Are the Women?" *Forbes Work in Progress*, June 20, 2012, http://www.forbes.com/sites/work-in-progress/2012/06/20/stem-fields-and-the-gender-gap-where-are-the-women/#3368918c33a9.

2. Joan C. Williams, "The 5 Biases Pushing Women out of STEM," *Harvard Business Review*, March 24, 2015, https://hbr.org/2015/03/the-5-biases-pushing-women-out-of-stem.

3. Ellen Pollack, "Why Are There Still So Few Women in Science?" *New York Times*, October 3, 2013, http://www.nytimes.com/2013/10/06/magazine/why-are-there-still-so-few-women-in-science.html?_r=0.

4. ADVANCE: Increasing the Participation and Advancement of Women in Academic Science and Engineering Careers, *National Science Foundation*, http://nsf.gov/funding/pgm_summ.jsp?pims_id=5383.

5. Invent It, Build It, *Society of Women Engineers*, http://societyofwomenengineers.swe.org/invent-it-build-it. Girls Who Code, http://girlswhocode.com/. EngineerGirl, http://www.engineergirl.org/. MakerGirl, http://makergirl.us/index.html. SciGirls, http://www.pbs.org/parents/scigirls/stemsantional-resources/tips-for-encouraging-girls-in-stem/.

6. Anna M. Lewis, *Women of Steel and Stone: 22 Inspirational Architects, Engineers, and Landscape Designers*, 2014, http://site.ebrary.com/id/10817604.

7. Betty Reynolds and Jill S. Tietjen, *Setting the Record Straight: An Introduction to the History and Evolution of Women's Professional Achievement* (Denver: White Apple Press, 2001).

8. Margaret W. Rossiter, *Women Scientists in America: Before Affirmative Action, 1940–1972* (Baltimore: Johns Hopkins University Press, 1995).

9. Martha Moore Trescott, *New Images, New Paths: A History of Women in Engineering in the United States 1850–1980* (Dallas: T & L Enterprises, 1996).

10. Margaret E. Layne, *Women in Engineering: Pioneers and Trailblazers* (Reston, Va: ASCE Press, 2009).

11. Margaret E. Layne, *Women in Engineering: Professional Life* (Reston, Va.: ASCE Press, 2009).

12. Amy Sue Bix, *Girls Coming to Tech! A History of American Engineering Education for Women*, 2013, http://public.eblib.com/choice/publicfullrecord.aspx?p=3339733.

13. S. Rugheimer, *Women in STEM Resources*, https://www.cfa.harvard.edu/~srugheimer/Women_in_STEM_Resources.html.

14. Martha Moore Trescott, "Women in the Intellectual Development of Engineering: A Study in Persistence and Systems Thought." In G. Kass-Simon and Patricia Farnes, eds. *Women of Science: Righting the Record* (Bloomington: Indiana University Press, 1990), 148.

15. Bix, *Girls Coming to Tech!*, 298.

16. Nupur Chaudhuri, Sherry J. Katz, and Mary Elizabeth Perry, *Contesting Archives: Finding Women in the Sources* (Urbana: University of Illinois Press, 2010).

Chapter 1. Engineers Who Happen to Be Women

1. First Report of the Board of Trustees, Record Series 1/1/802, University of Illinois Archives, 50.

2. Sixth Report of the Board of Trustees, Record Series 1/1/802, University of Illinois Archives, 27.

3. Enrollment Tables, 1885–1886, Record Series 25/3/810, University of Illinois Archives.

4. "Higher Education," *U.S. News & World Report*, http://colleges.usnews.rankingsandreviews.com/best-colleges/uiuc-1775.

5. "UIUC Student Enrollment," *Division of Management Information—University of Illinois*, http://dmi.illinois.edu/stuenr/#retention.

6. Julie Wurth, "Women Make Up a Quarter of UI College of Engineering's Freshman Class," (Champaign-Urbana) *News-Gazette*, August 8, 2016, http://www.news-gazette.com/news/local/2016-08-22/women-make-quarter-ui-college-engineerings-freshman-class.html.

7. *Catalogue. Illinois Industrial University*, 1874–1875, Record Series 25/3/801, University of Illinois Archives, 13.

8. Third Report of the Board of Trustees, Record Series 1/1/802, University of Illinois Archives, 27.

9. Graduate's Record, Mary Louise [*sic*] Page Folder, Alumni Morgue, 1882-, Record Series 26/4/1, University of Illinois Archives; Frank William Scott, ed. *The Alumni Record of the University of Illinois at Urbana; Including Annals of the University and Biographical Notices of the Members of the Faculties and of the Board of Trustees* (Urbana: University of Illinois, 1906), 40–41.

10. "Chapter Two—The Gregory Period, 1867, 1880—'Learning and Labor,'" Carl Stephens Manuscript History, Record Series 26/1/21, University of Illinois Archives, 2–28.

11. Tenth Report of the Board of Trustees, Record Series 1/1/802, University of Illinois Archives, 179; Student Ledger Books, 1879, Record Series 25/3/44, University of Illinois Archives.

12. Sarah Allaback, *The First American Women Architects* (Urbana: University of Illinois Press, 2008), 24.

13. Ibid.

14. Anna M. Lewis, *Women of Steel and Stone—22 Inspirational Architects, Engineers, and Landscape Designers* (Chicago: Chicago Review Press, 2014), 3.

15. "The Illinois School of Architecture: A History of Firsts," *The Illinois School of Architecture—University of Illinois*, http://www.arch.illinois.edu/welcome/history-school.

16. Allaback, *First American Women Architects*, 5.

17. Lewis, *Women of Steel and Stone*, 4–5.

18. Mary Louise [*sic*] Page Folder.

19. Louise J. Pellens Scrapbook, Record Series 41/20/86, University of Illinois Archives.

20. Twenty-Seventh Report of the Board of Trustees, Record Series 1/1/802, University of Illinois Archives, 235.

21. R. A. Kingery, R. D. Berg, and E. H. Schillinger, *Men and Ideas in Engineering—Twelve Histories from Illinois* (Urbana: University of Illinois Press, 1967), 1–2.

22. Lewis, *Women of Steel and Stone*, 40.

23. Allaback, *First American Women Architects*, 125–130.

24. "The Locomotive Laboratory, University of Illinois," *University of Illinois Bulletin*, Railway Engineering Department Folder, University Archives Reference File, Record Series 35/3/65, University of Illinois Archives.

25. Ira O. Baker, *A History of the College of Engineering of the University of Illinois 1868–1945* (Urbana: University of Illinois, 1947), 455; *Illinois Alumni News*, November 1962, Engineering, Electrical Folder, University Archives Reference File, Record Series 35/3/65, University of Illinois Archives.

26. "Physics Illinois Commends Its Exceptional Undergraduates," *Physics Illinois Bulletin* 1, No. 2 (2013): 49, http://physics.illinois.edu/alumni/Physics-Illinois-Bulletin-v1-No2.pdf.

27. Wilhelm Odelberg, *The Nobel Prizes 1977* (Stockholm: Nobel Foundation, 1978), http://www.nobelprize.org/nobel_prizes/medicine/laureates/1977/yalow-bio .html.

28. National Academy of Engineering, https://www.nae.edu/About/FAQ.aspx.

29. Society of Women Engineers, www.swe.org/webuiltthis.

30. Transactions of the Board of Trustees, 1930–32, Record Series 1/1/802, University of Illinois Archives, 186–187.

31. Angel Myers, "She's an Active Emeritus," (Champaign-Urbana) *News-Gazette*, July 16, 1978.

32. Amy S. Bix, "From 'Engineeresses' to 'Girl Engineers' to 'Good Engineers': A History of Women's U.S. Engineering Education," *NWSA Journal* 16, No. 1 (2004): 27–49.

33. Ibid.

34. "Coeds Study Engineering: 'Sole Girl in Class' Has Her Problems," *Daily Illini* (Urbana), October 13, 1955.

35. Betty Lou Bailey, interview by Deborah Rice, *Society of Women Engineers*, November 4, 2005.

36. Fran Myers, "Women Move into Engineering's Front Door, UI Meeting," (Champaign-Urbana) *News-Gazette*, October 28, 1962.

37. "Those Engineering Females," *Daily Illini* (Urbana), October 12, 1959.

38. William Wirtz, "Womanpower Boosted by Secretary of Labor," *Engineer* IV, No. 3 (Autumn 1963).

39. "Miss Lois Backer . . . What? An Engineer?" *Technograph*, March 1963.

40. "An ENGINEER Special Feature: The Woman Engineer," *Engineer* (Summer 1963).

41. College of Engineering 1966–1967 Annual Report, Record Series 11/1/3, University of Illinois Archives, 144.

42. College of Engineering 1967–1968 Annual Report, Record Series 11/1/3, University of Illinois Archives, 206.

43. Ibid., 191.

44. Martha Moore Trescott, "Women in the Intellectual Development of Engineering," in *Women of Science: Righting the Record*, eds. Gabriele Kass-Simon and Patricia Farnes (Bloomington: Indiana University Press, 1990).

45. "Women in Engineering It's Your Turn Now: A Career Counseling Conference," *University of Illinois* (Urbana), 1973.

46. "History of EOH," *Engineering Open House (EOH)*, http://eoh.ec.uiuc.edu/ #history.

47. "Engineering Open House Program—2001: An Engineering Odyssey," *College of Engineering, University of Illinois* (Urbana), 1975.

48. "Engineering—A Goal for Women," *Engineers' Council for Professional Development, Society of Women Engineers* (New York), 1979.

49. "UIUC Student Enrollment," *Division of Management Information—University of Illinois*, http://dmi.illinois.edu/stuenr/#retention.

50. Margaret Loftus, "Piercing the 20 Percent Ceiling," *Prism* (American Society for Engineering Education, February 2015), http://www.asee-prism.org/piercing-the-20 -percent-ceiling-feb/.

51. Women Engineering ProActive Network (WEPAN), https://www.wepan.org/.

Chapter 2. Early Inspiration

1. Amy Bix, *Girls Coming to Tech! A History of American Engineering Education for Women* (Cambridge: MIT Press, 2014), 19.

2. Ibid., 31.

3. Committee on the Education and Employment of Women i n Science and Engineering, *Climbing the Ladder: An Update on the Status of Doctoral Women Scientists and Engineers* (Washington, D.C.: National Academies Press, 1983), http://www.nap .edu/read/243/chapter/5#43.

4. Ibid.

5. Amy Bix, "From 'Engineeresses' to 'Girl Engineers' to 'Good Engineers': A History of Women's U.S. Engineering Education." In Margaret Layne, ed., *Women in Engineering: Pioneers and Trailblazers* (Reston, Va.: ASCE Press, 2009), 11–34.

6. Illinois Women Artists Project: Video on Louise Woodroofe, http://iwa.bradley. edu/multimedia/videos.

7. Transcript of Louise Woodroofe funeral service comments, courtesy of Barbara Schaede.

8. Jack S. Baker, Kathryn H. Anthony, Hub White, R. Alan Forrester, and Theodore Zernich, "A Tribute to Louise Woodroofe," Rickernotes, School of Architecture, University of Illinois, March 1996.

9. Fran Myers, "Broadwalk Tatler," (Champaign-Urbana) *News-Gazette*, April 3, 1960, 2.

10. Transcript of Louise Woodroofe funeral service comments.

11. Illinois Women Artists Project: Video on Louise Woodroofe.

12. Henry Fountain, "Putting Art in STEM," *New York Times*, October 31, 2014, http://www.nytimes.com/2014/11/02/education/edlife/putting-art-in-stem.html ?emc=edit_tnt_20141103&nlid=39451694&tntemail0=y&_r=3.

13. Lorella Jones, *An Introduction to Mathematical Methods of Physics* (Menlo Park, Calif.: Benjamin/Cummings Pub. Co., 1979).

14. Lorella Jones, letter to department head, Department of Physics archives, University of Illinois at Urbana-Champaign.

15. Lorella Jones, "Intellectual Contributions of Women in Physics." In Gabriele Kass-Simon and Patricia Farnes, eds. *Women of Science: Righting the Record* (Bloomington: Indiana University Press, 1990), 211.

16. Robert Delbourgo and Laura H. Greene, "Lorella M. Jones," *Physics Today* 48, No. 10 (October 2008): 90, http://physicstoday.scitation.org/doi/10.1063/1.2808224.

17. Judith Liebman, "Women in Civil Engineering Education," Proceedings of the ASCE Conference on Civil Engineering Education, Ohio State University, February 18–March 2, 1974.

18. Judith Liebman, "Women in Engineering at the University of Illinois in Urbana-Champaign," *IEEE Transactions on Education*, E18, No. 1 (February 1975): 47–49.

19. The Institute for Operations Research and the Management Sciences, https://www.informs.org/Recognize-Excellence/INFORMS-Prizes-Awards/Judith-Liebman-Award.

20. Margaret Rossiter, *Women Scientists in America: Before Affirmative Action 1940–1972* (Baltimore: Johns Hopkins Press, 1995).

21. Committee on the Education and Employment of Women in Science and Engineering, *Climbing the Ladder*.

22. Bix, "From 'Engineeresses' to 'Girl Engineers' to 'Good Engineers.'"

23. Betty Reynolds and Jill Tietjen, *Setting the Record Straight: An Introduction to the History and Evolution of Women's Professional Achievement* (Denver: White Apple Press, 2000), 71.

Chapter 3. Research Orientations

1. R. A. Kingery, R. D. Berg, and E. H. Schillinger, *Men and Ideas in Engineering—Twelve Histories from Illinois* (Urbana: University of Illinois Press, 1967), 1.

2. Ibid.

3. College of Engineering, University of Illinois at Urbana-Champaign, "Departmental Research," http://engineering.illinois.edu/research/departmental-research.html.

4. College of Engineering, University of Illinois at Urbana-Champaign, "Strategic Research Initiatives," http://engineering.illinois.edu/research/strategic-research-initiatives/.

5. This section has been adapted and updated by the coauthors with permission from Jeffrey Moore; August Schiess and Steve McGaughey are the original coauthors of "Collaboration Is the Key to Sottos' Success in Self-Healing Materials," the Beckman Institute for Advanced Science and Technology—University of Illinois, last modified March 4, 2014, http://beckman.illinois.edu/news/2014/03/collaboration-key-to-sottos-success-in-self-healing-materials.

6. This section has been adapted and updated by the coauthors with permission from Bill Bell; Mike Koon is the original author of "Hovakimyan Emerged from Cold War as Leader in Flight Control Safety," College of Engineering—University of Illinois, last modified May 20, 2015, http://mechanical.illinois.edu/news/hovakimyan-emerged-cold-war-leader-flight-control-safety.

7. Naira Hovakimyan and Chengyu Cao, *L1 Adaptive Control Theory: Guaranteed Robustness with Fast Adaptation (Advances in Design and Control)* (New Zealand: Society for Industrial and Applied Mathematics, 2010).

8. Clair Sullivan, in discussion with the author, December 12, 2015.

9. Ibid.

10. Ibid.

11. Ibid.

12. Susan Mumm, "Sullivan Wins ANS Oestmann Achievement Award," Department of Nuclear, Plasma, and Radiological Engineering—University of Illinois, last modified September 2, 2015, http://npre.illinois.edu/news/sullivan-wins-ans-oestmann -achievement-award.

13. Clair Sullivan interview.

14. Ibid.

15. Mumm, "Sullivan Wins ANS Oestmann Achievement Award."

16. "Sullivan Wins DARPA Young Faculty Award," Department of Nuclear, Plasma, and Radiological Engineering—University of Illinois, last modified September 24, 2014, http://npre.illinois.edu/news/sullivan-wins-darpa-young-faculty-award.

17. Mumm, "Sullivan Wins ANS Oestmann Achievement Award."

18. Ibid.

Chapter 4. Relentless Innovators

1. Jeff Dyer, Hal B. Gregersen, and Clayton M. Christensen, *The Innovator's DNA: Mastering the Five Skills of Disruptive Innovators* (Boston: Harvard Business Press, 2011).

2. Katherine W. Phillips, "How Diversity Makes Us Smarter," *Scientific American*, October 1, 2014, http://www.scientificamerican.com/article/how-diversity-makes -us-smarter/.

3. Joel Lyons, "Google to Spend $150 Million on Diversity Initiatives in 2015," *Black Enterprise*, May 7, 2015, http://www.blackenterprise.com/technology/google-spend-150 -million-diversity-initiatives-2015/.

4. "Illinois Receives Higher Education Excellence in Diversity Award," University of Illinois at Urbana-Champaign, http://diversity.illinois.edu/doc/UIUC_HEED_ Award_2016.pdf.

5. "Fiscal Year 2015 Annual Report," *University of Illinois, Office of Technology Management*, http://otm.illinois.edu/sites/default/files/Fiscal%202015%20annual%20report .pdf.

6. "Top Innovative Universities—2015," Reuters, http://www.reuters.com/most -innovative-universities/profile.

7. Jennifer Pocock, "Going for Broke," *ASEE Prism*, December 2015.

8. Julia C. Keller, "Looking at the Big Picture," *IEEE Women in Engineering Magazine* (Winter 2007/2008): 25–27.

9. Paul Lasewicz, "IBM 'Women in Technology' Archive," Oral History Interview, Dr. Joan L. Mitchell, April 2, 2003, https://www-3.ibm.com/ibm/history/witexhibit/ pdf/mitchell_history.pdf.

10. "Joan Mitchell," https://www-03.ibm.com/ibm/history/witexhibit/wit_fellows _mitchell.html.

11. "Joan Mitchell," *Engineering at Illinois*, https://engineering.illinois.edu/engage/distinguished-alumni-and-friends/hall-of-fame/2011/joan-mitchell.

12. Susan Fry, "Spotlight: Joan Mitchell, '69—Digital Data Pioneer," *Stanford Alumni*, https://alumni.stanford.edu/get/page/magazine/article/?article_id=28541.

13. Lasewicz, "IBM 'Women in Technology' Archive."

14. "Distinguished Alumni Awards—Joan L. Mitchell, Engineering at Illinois," https://engineering.illinois.edu/engage/distinguished-alumni-and-friends/distinguished/article/5747.

15. "IBM Fellow Shares Life in Industry with Students," *Physics Illinois News 2005*, http://physics.illinois.edu/alumni/newsletters/PIN-2005-No1.pdf.

16. "Jennifer A. Lewis," *Harvard—John A. Paulson School of Engineering and Applied Sciences*, https://www.seas.harvard.edu/directory/jalewis.

17. Jennifer Lewis in discussion via email with the author, November 21, 2015.

18. Ibid.

19. Ibid.

20. Jerome Groopman, "Print Thyself," *New Yorker*, last modified November 24, 2014, http://www.newyorker.com/magazine/2014/11/24/print-thyself.

21. Lewis discussion with author.

22. Analisa Russo in discussion with the author, October 20, 2015.

23. Ibid.

24. Lewis discussion with author.

25. Ibid.

26. "A World Disrupted: The Leading Global Thinkers of 2014," *Foreign Policy*, http://globalthinkers.foreignpolicy.com/.

27. "The Most Creative People in Business 2015," *Fast Company*, https://www.fastcompany.com/most-creative-people/2015.

28. "Jennifer A. Lewis."

29. Lewis discussion with author.

30. Ibid.

31. Ibid.

32. Ibid.

33. This section has been adapted and updated by the coauthors with permission from Rakesh Nagi; Emily Scott is the original author of "Amy Doroff BSGE 2015," Industrial And Enterprise Systems Engineering—University of Illinois, last modified January 14, 2016, http://ise.illinois.edu/newsroom/article/amy-doroff-bsge-2015.

34. Amy Doroff in discussion with the author, October 11, 2015.

35. "BS Degree in Innovation, Leadership & Engineering Entrepreneurship (ILEE)," *Technology Entrepreneurship Center—Engineering at Illinois*, http://tec.illinois.edu/academics/degree.

36. Mike Koon, "IBHE Approves Innovation, Leadership and Engineering Entrepreneurship Degree," last modified September 27, 2016, http://engineering.illinois.edu/news/article/19383.

37. "First Class of Engineering Faculty Entrepreneurial Fellows Announced," *Engineering at Illinois*, last modified July 30, 2015, http://engineering.illinois.edu/news/article/11533.

38. Koon, "IBHE Approves Innovation."

Chapter 5. Acclaim in Bioengineering and Medicine

1. B. Yoder, "Engineering by the Numbers," American Society for Engineering Education, 2017, https://www.asee.org/documents/papers-and-publications/publications/college-profiles/16Profile-Front-Section.pdf.

2. L. Nilsson, "How to Attract Female Engineers," The Opinion Page, *New York Times*, April 27, 2015, https://www.nytimes.com/2015/04/27/opinion/how-to-attract -female-engineers.html?_r=0.

3. Wilhelm Odelberg, ed., *Les Prix Nobel*, Rosalyn Yalow, Nobel Prize biography (Stockholm, Sweden: Nobel Foundation, 1978).

4. Sharon Bertsch McGrayne, *Nobel Prize Women in Science: Their Lives, Struggles, and Momentous Discoveries: Second Edition* (Washington, D.C: Joseph Henry Press, 1998), https://doi.org/10.17226/10016.

5. Odelberg, *Les Prix Nobel.*

6. *Discover*, April 1982, 82.

7. McGrayne, *Nobel Prize Women in Science.*

8. Ibid.

9. Ibid.

10. Department of Physics records, University of Illinois at Urbana-Champaign, Urbana, Illinois.

11. Odelberg, *Les Prix Nobel.*

12. Dennis Overbye, "Rosalyn Yalow: Lady Laureate of the Bronx," *Discover*, June 1982.

13. Odelberg, *Les Prix Nobel.*

14. Ibid.

15. Rosalyn Yalow, "A Physicist in Biomedical Investigation," American Institute of Physics News Release, Joint Annual Meeting of the American Physical Society and the American Association of Physics Teachers, January 30, 1979. Unpublished.

16. Odelberg, *Les Prix Nobel.*

17. Yalow, "Physicist in Biomedical Investigation."

18. *Discover*, April 1982.

19. McGrayne, *Nobel Prize Women in Science.*

20. R. S. Yalow and S. A. Berson, "Some Applications of Isotope Dilution Techniques," *American Journal of Roentgenology, Radium Therapy, and Nuclear Medicine* 75 (1956): 1059–1067.

21. McGrayne, *Nobel Prize Women in Science.*

22. Ibid.

23. John Boslough, "Rosalyn Yalow: Revolutionizing Medical Research," *U.S. News & World Report*, September 8, 1980, 66–67.

24. McGrayne, *Nobel Prize Women in Science*.

25. Overbye, "Rosalyn Yalow."

26. McGrayne, *Nobel Prize Women in Science*.

27. *Discover* magazine, ibid.

28. Overbye, "Rosalyn Yalow."

29. Rosalyn Yalow's speech at the Nobel Banquet, December 10, 1977, http://www.nobelprize.org/nobel_prizes/medicine/laureates/1977/yalow-speech.html.

30. Ibid.

31. Ibid.

32. This section has been adapted and updated by the coauthors with permission from Michael Insana; Claire Sturgeon is the original author of "Bioengineering's Princess Imoukhuede and Team Working on Developing Personalized Cancer Treatment Approach," Department of Bioengineering—University of Illinois, last modified October 29, 2014, http://bioengineering.illinois.edu/news/bioengineerings-princess-imoukhuede-and-team-working-developing-personalized-cancer-treatment-a.

33. Susan McKenna, Department of Bioengineering, personal communication.

34. Rosalyn Yalow's speech.

Chapter 6. Touching the Sky

1. "History of Aerospace Engineering at Illinois," *Aerospace Engineering University of Illinois at Urbana-Champaign*, 2015, http://aerospace.illinois.edu/about-us/history-aerospace-engineering-illinois.

2. Juergen Teller, "One Giant Leap for Womankind," *Walter P. Ruether Library*, July 6, 2011, http://reuther.wayne.edu/node/7948.

3. "The 50th Anniversary of the Space Age," *National Aeronautics and Space Administration*, http://www.nasa.gov/externalflash/SpaceAge/.

4. Teller, "One Giant Leap for Womankind."

5. "Social, Cultural, and Educational Legacies," *NASA*, http://www.nasa.gov/centers/johnson/pdf/584743main_Wings-ch6a-pgs459–469.pdf.

6. Martha Trescott, *New Images, New Paths: A History of Women in Engineering in the United States 1850–1980* (Dallas: T&L Enterprises, 1996), 67–69.

7. Ibid.

8. Ibid.

9. Ibid.

10. Barbara Crawford Johnson, interviewed by Lauren Kata, *Profiles of SWE Pioneers Oral History Project, Walter P. Reuther Library and Archives of Labor and Urban Affairs*, Wayne State University, May 3, 2003.

11. Ibid.

12. Trescott, *New Images, New Paths*, 142.

13. Johnson interview.

14. Ibid.

15. Ibid.

16. *Mach number* is the ratio of an object's speed to the speed of sound (approximately 760 miles per hour). A number greater than 5 indicates that the object is hypersonic as it is traveling at a speed five times faster than the speed of sound. Numbers between 1 and 5 indicate supersonic conditions, and numbers less than 1 indicate subsonic conditions. "Mach Number," *NASA*, https://www.grc.nasa.gov/www/k-12/airplane/mach.html.

17. Ibid.

18. Ibid.

19. Ibid.

20. The Dyna-Soar program focused on the development of a hypersonic (Mach number greater than 5) glider that was a precursor to the space shuttle. It was designed to be launched on a missile, orbit the Earth, and land on a runway. Its travel was considered dynamic soaring, hence the name. "When the Dyna-Soar Went Extinct," *Seeker*, December 12, 2012, http://news.discovery.com/space/history-of-space/death-of-the-dyna-soar-121212.htm.

21. The concept of lunar orbital rendezvous allowed for the completion of the first lunar mission, which fired "an assembly of three spacecraft into Earth orbit on top of a single powerful (three-stage) rocket. This assembly included: one, a mother ship, or command module; two, a service module containing the fuel cells, attitude control system, and main propulsion system; and three, a small lunar lander or excursion module. Once in Earth orbit, the last stage of the rocket would fire, boosting the Apollo spacecraft with its crew of three men in to its flight trajectory to the moon." "The Rendezvous That Was Almost Missed: Lunar Orbit Rendezvous and the Apollo Program," *NASA*, December 1992, http://www.nasa.gov/centers/langley/news/factsheets/Rendezvous.html.

22. Ibid.

23. Trescott, *New Images, New Paths*, 144.

24. Johnson interview.

25. Ibid.

26. Trescott, *New Images, New Paths*, 346–349.

27. Ibid.

28. "Barbara Crawford Johnson," *Engineering at Illinois*, 1975, http://engineering.illinois.edu/engage/distinguished-alumni-and-friends/distinguished/article/5703.

29. Ibid.

30. Johnson interview.

31. Victoria Coverstone in discussion with the author, April 3, 2015.

32. "Victoria Coverstone," *EngineerGirl*, August 28, 2007, http://www.engineergirl.org/Engineers/interviews/7103.aspx.

33. Ibid.

34. Coverstone in discussion with the author, June 8, 2015.

35. Ibid., April 3, 2015.

36. Ibid.

37. Victoria Coverstone-Carroll, *Personal Diary*, 1994–2004, Unpublished Manuscript.

38. Ibid.

39. "Victoria Coverstone," *EngineerGirl*.

40. Ibid.

41. "Victoria Lynn Coverstone," *Aerospace Engineering—University of Illinois*, http://aerospace.illinois.edu/directory/profile/vcc.

42. Susan Mumm, "Coverstone Named AIAA Fellow," *Aerospace Engineering University of Illinois at Urbana-Champaign*, January 8, 2013, http://aerospace.illinois.edu/news/coverstone-named-aiaa-fellow.

43. Coverstone in discussion with the author, April 3, 2015.

44. Mumm, "Coverstone Named AIAA Fellow."

45. Coverstone in discussion with the author, June 10, 2015.

46. Ibid., July 12, 2015.

47. Brian Yoder, "Engineering by the Numbers," *American Society for Engineering Education*, https://www.asee.org/papers-and-publications/publications/college-profiles/15EngineeringbytheNumbersPart1.pdf.

Chapter 7. Women's Work

1. *ILLIAC Programming: A Guide to the Preparation of Problems for Solution by the University of Illinois Digital Computer, Digital Computer Laboratory*, Record Series 11/15/804, University of Illinois Archives.

2. Bethany Anderson, "The Birth of the Computer Age at Illinois," *University of Illinois Archives*, last modified September 23, 2013, http://archives.library.illinois.edu/blog/birth-of-the-computer-age/.

3. Heidi Hartmann, *Computer Chips and Paper Clips: Technology and Women's Employment, Volume II—Case Studies and Policy Perspectives* (Washington, D.C.: National Academy Press, 1987), 139.

4. *ILLIAC Programming*.

5. Hartmann, *Computer Chips and Paper Clips*, 139.

6. Amr Elnashai and William Hall, *Leadership and Legacy: A History of Civil and Environmental Engineering at Illinois* (Champaign: University of Illinois Department of Civil & Environmental Engineering, 2011).

7. "In Memoriam—Nancy B. Brooks," *CEE Magazine* (Summer 2009): 38.

8. "Our Women in Engineering—Nancy B. Brooks," *CEE Magazine* (Spring/Summer 2003): 9.

9. Ibid.

10. Ibid.

11. William Hall, personal email correspondence, February 12, 2015.

12. Nancy Brooks, *Development of Procedures for Rapid Computation of Dynamic Structural Response—Final Report for the Period July 1, 1953 to June 30, 1954* (Champaign: University of Illinois—Civil Engineering Studies, 1954). Nancy Brooks and Dolly Marsh, *Development of Procedures for Rapid Computation of Dynamic Structural Response—Final Report for the Period July 1, 1954 to June 30, 1955* (Champaign: University of Illinois—Civil Engineering Studies, 1955).

13. "Our Women in Engineering—Nancy B. Brooks."

14. Ibid.

15. "In Memoriam—Nancy B. Brooks."

16. This section has been adapted and updated by the coauthors with permission from William Sanders; Jonathan Damery is the original author of "Field Report—Hillery Hunter," *Resonance* (Fall 2014): 30–31. Department of Electrical and Computer Engineering—University of Illinois.

17. Ibid.

18. Julie Wurth, "Google's 'Security Princess' a Hacker's Nightmare," (Champaign-Urbana) *News-Gazette*, October 24, 2014, http://www.news-gazette.com/news/local/2014-10-24/googles-security-princess-hackers-nightmare.html.

19. Clare Malone, "Meet Google's Security Princess," *ELLE*, July 8, 2014, http://www.elle.com/culture/tech/a14652/google-parisa-tabriz-profile/.

20. Parisa Tabriz in discussion with the author, October 23, 2014.

21. Ibid.

22. Ibid.

23. Ibid.

24. Brian Yoder, "Engineering by the Numbers," *American Society for Engineering Education*, https://www.asee.org/papers-and-publications/publications/college-profiles/15EngineeringbytheNumbersPart1.pdf.

25. Sadie Plant, "The Future Looms: Weaving Women and Cybernetics," *Body & Society* 1, No. 3–4 (November 1995): 45–64. Virginia Postrel, "Losing the Thread," *Aeon Magazine*, June 5, 2015, https://acon.co/essays/how-textiles-repeatedly-revolutionised-human-technology. Yasmin Kafai, Kristin Searle, Crîstobal Martinez, and Bryan Brayboy, "Ethnocomputing with Electronic Textiles: Culturally Responsive Open Design to Broaden Participation in Computing in American Indian Youth and Communities," *Proceedings of the 45th ACM Technical Symposium on Computer Science Education* (2014): 241–246.

26. The most distinctive garment worn by Mayan women is the *huipil*, a blouse usually made of two panels of cotton fabric woven on a backstrap loom and joined along the side selvages with the seams worn vertically. "Mayan Costumes of Guatemala," Museum of Anthropology, College of Arts and Science, University of Missouri, https://anthromuseum.missouri.edu/minigalleries/guatemalatextiles/intro.shtml.

27. Julia Hendon, "Textile Production as Craft in Mesoamerica Time, Labor and Knowledge," *Journal of Social Archaeology* 6, No. 3 (October 2006): 354–378.

28. Ron Eglash, *The Fractals at the Heart of African Designs*, video, 16:57, June 2007, http://www.ted.com/talks/ron_eglash_on_african_fractals?language=en.

29. *Recursion* is defined in computer science as a method where the solution to a problem depends on solutions to smaller instances of the same problem (as opposed to iteration). Ronald Grahm, Donald Knuth, and Oren Patashnik, *Concrete Mathematics* (Boston: Addison-Wesley Publishing Company, 1990), Chapter 1: "Recurrent Problems."

30. Optimization problems involve finding the best or "optimal" solution among all potential solutions. Paul E. Black, entry for data structure in *Dictionary of Algorithms and Data Structures*, *U.S. National Institute of Standards and Technology*, https://xlinux.nist.gov/dads/.

31. The defining of data structures involves organizing data in a particular way so it can be used efficiently. Black, entry for data structure in *Dictionary of Algorithms and Data Structures*.

32. Ron Eglash, *African Fractals: Modern Computing and Indigenous Design 3rd Edition* (New Brunswick: Rutgers University Press, 1999). Helen Verran, *Science and an African Logic* (Chicago: University of Chicago Press, 2001). Paulus Gerdes, *Women, Art and Geometry in Southern Africa* (Trenton: Africa World Press, 1998).

Chapter 8. Global Challenges

1. "Global Challenges," National Academies, http://sites.nationalacademies.org/International/international_052200.

2. "Black Carbon," Environmental Protection Agency, https://www3.epa.gov/blackcarbon/.

Chapter 9. Do Engineers Have to Engineer?

1. Elsie Eaves, "Civil Engineering," in *An Outline of Careers for Women 1928*, ed. Doris E. Fleischman (New York: Double Day, Doran and Company, 1928), 151.

2. Dinesh Paliwal, "Engineering Is the New Liberal Arts," *Forbes.com*, last modified February 24, 2016, http://onforb.es/1S1aco8.

3. Beryl Lieff Benderly, "Checkered Careers," *Asee-Prism.org*, last modified January 2015, http://www.asee-prism.org/checkered-careers-jan/.

4. This section has been adapted and updated by the coauthors with permission from Paul Kenis; Christine des Garennes is the original author of "Kathryn 'Kit' Gordon, BS '83—Tech Innovator and Watershed Protector," *Chemical and Biomolecular Engineering at Illinois* (Spring/Summer 2016): 23, last modified August 2, 2016, Department of Chemical and Biomolecular Engineering—University of Illinois at Urbana-Champaign, http://chbe.dev.engr.illinois.edu/wp-content/uploads/2016/03/ChBE-Summer2016_web_pages.pdf.

5. "Semiconductor Antifuse Structure and Method," https://www.google.ch/patents/EP0414361A2?cl=en.

6. "Biography—Laurie Morvan," *Laurie Morvan Band*, http://www.lauriemorvan.com/images/LaurieMorvan-Bio.pdf.

7. "Where Are the Girls with Guitars?" *Resonance* (Winter 2009–2010), https://www.ece.illinois.edu/newsroom/resonance/pdf/2009_Fall.pdf.

8. "Laurie Morvan Band: Nothing but the Blues," *Laurie Morvan Band*, http://www.lauriemorvan.com/reviews/2010/10_big-ten-network-tv-special.html.

9. Melissa Merli, "Former UI Volleyball Hitter Laurie Morvan Making Sweet Music," (Champaign-Urbana) *News-Gazette*, last modified July 25, 2008, http://www.news-gazette.com/sports/illini-sports/volleyball/2008-07-25/former-ui-volleyball-hitter-laurie-morvan-making-sweet-mu.

10. "Laurie Morvan Is the New Blues," *Local Limit*, http://www.dabelly.com/columns/lint88.htm.

11. "Music," *Laurie Morvan Band*, http://www.lauriemorvan.com/music.asp.

Chapter 10. Mentors and Mentoring

1. Remarks by the President on the "Educate to Innovate" Campaign and Science Teaching and Mentoring Awards, *White House*, https://www.whitehouse.gov/the-press-office/remarks-president-educate-innovate-campaign-and-science-teaching-and-mentoring-awar.

2. Naomi Chesler and Mark Chesler, "Gender-Informed Mentoring Strategies for Women Engineering Scholars: On Establishing a Caring Community," *Journal of Engineering Education 91* (2002): 49–55, doi: 10.1002/j.2168-9830.2002.tb00672.x. Sylvia Thomas, "Mentoring Women STEM Faculty: Key Strategies for Career and Institutional Progression," in *Alliances for Advancing Academic Women*, eds. Penny Gilmer, Berren Tansel, and Michellel Hughes Miller (Rotterdam: Sense Publishers, 2014), 147–164.

3. Cara Poor and Shane Brown, "Increasing Retention of Women in Engineering at WSU: A Model for a Women's Mentoring Program," *College Student Journal 47*, No. 3 (2013): 421–428.

4. Becky Packard, *Successful STEM Mentoring Initiatives for Underrepresented Students: A Research-Based Guide for Faculty and Administrators* (Sterling, Va.: Stylus Publishing, 2015).

5. "Million Women Mentors Program," https://www.millionwomenmentors.org/#home.

6. "Tech Women," https://www.techwomen.org/.

7. Joi-Lynn Mondisa, *In the Mentor's Mind: Examining the Experiences of African-American STEM Mentors in Higher Education* (Order No. 3734512), 2015. Available from Dissertations & Theses @ CIC Institutions; ProQuest Dissertations & Theses Full Text; ProQuest Dissertations & Theses Global (1736117866), http://search.proquest.com/docview/1736117866?accountid=14553.

8. Judith Liebman, "Women in Engineering at the University of Illinois in Urbana-Champaign," *IEEE Transactions on Education* E18, No. 1 (1975): 47–49.

9. Martha Moore Trescott, *New Images, New Paths: A History of Women in Engineering in the United States, 1850–1980* (Dallas: T&L Enterprises, 1996).

10. Angel Myers, "She's an Active Emeritus," (Champaign-Urbana) *News-Gazette*, July 16, 1978.

11. Tom Moone, "Celebrating 10 Years of the Siebel Center," *CS @ ILLINOIS*, April 24, 2014.

12. "Tom Siebel Video Tribute to Geneva Belford on the Occasion of her 2012 CS Distinguished Service Award," https://www.youtube.com/watch?v=YtjqC8sGSZI.

13. Geneva Grosz Belford obituary, (Champaign-Urbana) *News-Gazette*, March 9, 2014.

14. Society of Women Engineers, *Be That Engineer: Inspiration and Insight from Accomplished Women Engineers* (Chicago: Society of Women Engineers: 2014).

Chapter 11. Inspiring the Next Generation

1. Lillian M. Gilbreth, "Industrial Engineering," in *An Outline of Careers for Women 1928*, ed. Doris E. Fleischman (New York: Double Day, Doran and Company, 1928), 167.

2. W. L. Everitt, "A Message from the Dean," *Technograph*, December 1950, 1.

3. Amy Bix, *Girls Coming to Tech!* (Cambridge, Mass.: MIT Press, 2013), 213.

4. "Engineers Honor High School Girls—Name 'Jr. Queens' for Open House," *Daily Illini* (Urbana), March 14, 1962.

5. Bix, *Girls Coming to Tech!* 268.

6. Betty Lou Bailey, interview by Deborah Rice, *Betty Lou Bailey*, Society of Women Engineers, Anaheim, California, November 4, 2005.

7. Betty Lou Bailey, "Women Can Be Engineers" (presentation, University of Illinois College of Engineering, Urbana, October 27, 1962).

8. Bailey, interview.

9. Bailey, presentation.

10. Bailey, interview.

11. Ibid.

12. Ibid.

13. Ibid.

14. Bailey, presentation.

15. Bailey, interview.

16. Ibid.

17. Ibid.

18. Ibid.

19. Bailey, presentation.

20. Bailey, interview.

21. Boyd Stevens, "Question-Air: Girls, How About Engineering Career?" *Free Lance-Star*, September 2, 1967.

22. "MechSE Distinquished Alumni," *Department of Mechanical Science and Engineering*, https://mechanical.illinois.edu/alumni/alumni-awards/mechse-distinguished-alumni.

23. Mal Provost, "Memories of Betty Lou Bailey," *Adirondack Mountain Club—Schenectady Chapter*, last modified July 07, 2015, http://www.adk-schenectady.org/chapter_history#TOC-Memories-of-Betty-Lou-Bailey.

24. "Notable Members," *Philadelphia Society of Women Engineers*, http://philadelphia.swe.org/hall-of-fame-a—l.html.

25. Brian Yoder, "Engineering by the Numbers," *American Society for Engineering Education*, https://www.asee.org/papers-and-publications/publications/college-profiles/15EngineeringbytheNumbersPart1.pdf.

26. E. Alpay, A. Hari, M. Kambouri, and A. Ahearn, "Gender Issues in the University Research Environment," *European Journal of Engineering Education* 35, No. 2 (2010): 135–145, doi: 10.1080/03043790903497302.

27. Yoder, "Engineering by the Numbers."

Chapter 12. Foundations for the Future

1. CMG Worldwide, "Hedy Lamarr: The Official Site," http://www.hedylamarr.com/about/biography.html. Wikipedia, "Hedy Lamarr," https://en.wikipedia.org/wiki/Hedy_Lamarr.

2. *Encyclopedia Britannica*, "Emily Warren Roebling," https://www.britannica.com/biography/Emily-Warren-Roebling. Wikipedia, "Emily Warren Roebling," https://en.wikipedia.org/wiki/Emily_Warren_Roebling.

3. *Encyclopedia Britannica*, "Stephanie Kwolek," https://www.britannica.com/biography/Stephanie-Kwolek.

4. Wikipedia, "Mary Anderson (Inventor)," https://en.wikipedia.org/wiki/Mary_Anderson_(inventor). Google, Patents, "Window Cleaning Device," https://www.google.com/patents/US743801. U.S. Patent Office, *Official Gazette of the US Patent Office, Vol. 107, No. 1*, http://tinyurl.com/h93s69j.

5. B. Yoder, "Engineering by the Numbers," *American Society for Engineering Education*, 2017, https://www.asee.org/documents/papers-and-publications/publications/college-profiles/16Profile-Front-Section.pdf.

INDEX

LAURA D. HAHN is the director of the Academy for Excellence
in Engineering Education at the University of Illinois
Urbana-Champaign.

ANGELA S. WOLTERS is the director of Women in Engineering
at the University of Illinois Urbana-Champaign.

The University of Illinois Press
is a founding member of the
Association of American University Presses.

Composed in 10.25/13 Marat Pro
with Trade Gothic Condensed display
by Kirsten Dennison
at the University of Illinois Press
Cover designed by Jennifer S. Fisher
Cover illustration: *The Quintessential Engineer*
by Julie Rotblatt-Amrany, Engineering Quad,
University of Illinois at Urbana-Champaign.
Photo courtesy of Thompson-McClellan.
Manufactured by Versa Press, Inc.

University of Illinois Press
1325 South Oak Street
Champaign, IL 61820-6903
www.press.uillinois.edu